EARLY TECHNOLOGIES

UNDENA PUBLICATIONS
MALIBU 1979

Invited Lectures on the Middle East at the University of Texas at Austin

Volume Three

Early Technologies

edited by Denise Schmandt-Besserat

Library of Congress Card Number: 78-59574
ISBN: 0-89003-031-6 (hard cover)
0-89003-032-4 (paper cover)

Undena Publications, P.O. Box 97, Malibu, Ca. 90265

Table of Contents

List of Illustrations

Acknowledgements

The exhibit "Locks from Iran: Pre-Islamic to 20th Century" was shown in Austin at the University of Texas Art Museum during the fall of 1976. In conjunction with the exhibit, the Center for Middle Eastern Studies and the Art Museum co-sponsored a symposium on early technologies which brought together scholars from the fields of art, anthropology, history of science, and history. The events took place at the Academic Center of the University of Texas at Austin on October 25-27, 1976.

I would like to thank Paul W. English, Director of the Center for Middle Eastern Studies, and Donald B. Goodall, Director of Collections, for their generous help in making the project possible. The symposium was funded by several units of the University of Texas and I would like to express my appreciation to the individuals who supported the project; in particular Irwin C. Lieb, Vice President for Academic Affairs, Peter Garvie, Dean of the College of Fine Arts, Ralph J. Kaufman, Chairman of Comparative Studies, John B. Cornell, Chairman, Department of Anthropology, and Robert A. Fernea, Chairman, Public Lecture Committee. I would like to thank all those who took part in the organization of the project and particularly Theodore A. Wertime who helped considerably during the planning phase; Marian B. Davis who gracefully opened the events with welcoming words; and the chairmen of the various sessions who directed the lively and insightful discussions following each presentation: Dee Ann Story, Director of the Anthropological Laboratory, E. Mott Davis, Director of the Radio Carbon Laboratory, and John A. Wertime, collector. I am indebted to all participants: S.M. Alexander, Jeremiah Epstein, C. C. Lamberg-Karlovsky, Harry Miller, James A. Neely, Bert S. Hall, Cyril S. Smith, Theodore A. Wertime, and John A. Williams; and in particular to Derek de Solla Price, Avalon Professor of History of Science at Yale University, for the highly stimulating keynote address he delivered. I would like to thank the distinguished visitors who shared their knowledge in the various discussions and, in particular James D. Muhly, Professor of Ancient History, the University of Pennsylvania, and Frank Hole, Professor of Anthropology, Rice University. Finally, I wish to thank very specially, all the staff of the Center for Middle Eastern Studies for the innumerable tasks they performed during all phases of the project, and in particular, Scott Lubeck, Editor; and Sue Ellen Jeffers, Secretary.

Denise Schmandt-Besserat
Assistant Professor of Art and Archaeology
Coordinator of the Symposium

Preface

by

Cyril Stanley Smith

The organizers of this conference tell me that it is the first conference in the English-speaking world to be devoted exclusively to its topic as distinct from sections and sessions on technology within conferences covering much broader areas of archaeological or historical interest. How astonishing that our technology-dominated society should wait until 1976 for such a view of its origins! The reason for the neglect, which has allowed widespread misunderstanding to persist, seems to lie partly in the fact that the sources for the study of technology are rarely the verbal ones beloved by the literate scholar. Moreover, as Derek Price pointed out in his opening lecture, the developments themselves do not show the search for inner logic that has driven both science itself and the study of its history. Yet the very irrationality of the origins of technology and the generally unconscious beginnings of its interaction with social purposes make its history a more "human" one than intellectual history *per se*. It is far more easily relatable to the concerns of the average person than the political history that has been so intensively studied.

If one extrapolates from the way in which interest has grown during the last decade or two, the history of technology will before long become a respectable component of the historian's concern, with disciplinary boundaries, well defined sub-fields and high professional standards. This has not yet happened, however, and alongside some of the world's best scholars the field still has room for individual enthusiasts and well-informed but undisciplined amateurs, and it displays an uneasy balance between too-fine detail and too-broad generalizations. The diversity of approach among the papers in the present volume does not reflect a well-structured discipline, but it certainly makes for exciting reading!

Three things in particular stand out: first, that the earliest technology was the modification of the properties of materials by the use of fire; second, that the first mechanisms were models of theories, were in fact theories themselves, experimentally discovered and mechanically demonstrated; and third, that art played a role both in the initial discovery of the diversity and properties of matter, and, much later, in perspective drawing, allowing quick tests of newly-conceived mechanical interactions.

There is little support for the popular beliefs that technology is applied science and that science is done because of its potential applicability. To quote Derek Price: Science "starts with astronomy not because it is useful but because of the weird properties of the human mind and of nature. . . . Science, in fact, was not useful for anything, not technology nor any application, until the very end of the 18th century." He suggests that the main function of instruments was to reveal new worlds for philosophy, new aspects of nature that man had never seen because nature had never created them. The same viewpoint can be extended to the period even before astronomy began, to the beginnings of chemistry and to the beginnings of technology itself. Note the role of the decorative arts in the early pyrotechnologies discussed by Wertime—the use of fire in the modification of iron oxide minerals for use as pigments about 50,000 years ago, and the heat-treatment of flint in Solutrean France (about 20,000 B.C.) which, to quote Epstein, was "most useful in the superb mastery of flint that is required to produce an object whose beauty transcends its functional requirements." And all of the more mature technologies—plaster pots, glazes and glass; metals with the transition from discoveries for use in jewelry and sculpture and the tools to make them—show that the crafts first uncovered aspects of nature upon which philosophies were later to be built. No, technology is not applied science. Rather, as we are only now coming to see, it has usually been the other way around. The beginnings of technology and the beginnings of science both depend on human curiosity, a search for aesthetic satisfaction that involves the whole mind and body, the sensual and aesthetic reactions of man as well as his intellectual abilities. The contributions of science to understanding have been enormous and beautiful, but the first perception of an interesting area to which science can be applied have usually not come from theory. Philosophers (and perhaps scientists seeking patronage?) have turned things upside down and have emphasized the logical postlude over the aesthetic precedent. Papers such as those at the present conference and the more detailed scrutiny of the technologists' experience that is possible in later periods of history are beginning to show what the humanist has always known, that man is greater than his analyzable parts, and to distinguish those conditions under which change moves unpredictably from the part upward from those in which it moves almost predictably from the environment downward.

The selfless devotion of the artist is highly praised in our society, and most people somehow feel better about technology when told of its probable origin in the search for decorative effects. There is, however, an interesting and perhaps even alarming angle to this. Murder has been committed to acquire beautiful things, though rarely by their makers. All man's attributes, noble and base alike, have to be judged in terms of their contribution to the survival of his species. Today, although the socially-necessary entrepreneur seems to be mainly motivated by anticipation of monetarily-measurable reward, those who originate technology still do so quite as much for the rich personal satisfaction of their work as for the hope that they will be paid for it. An article in the *Bulletin of the Atomic Scientists* (October 1977) that arrived when I was writing this suggests that the absurd excess of militarism in our society is largely the result of the desire to play: The military is a gigantic communal toy-owning organization, providing the technophiles among us with the opportunity to play with the neatest and newest toys. Is it perhaps the very same drive toward aesthetic satisfaction that led to the discovery of all "good" technology when it was individual and unorganized that, when reinforced at another level by socialization and organization to form what Derek Price dubs high technology, loses its human roots and forms what Lewis Mumford has called the Megamachine with all its ominous portents? We desperately need to discover the complexly-balanced aesthetics of larger systems and to recognize the distinction between the social function of the initially unique creative act and that of repetitive production.

The topics covered in this conference show that the conditions for discovery or prime invention are different from those surrounding their development and acceptance into the social system. The conditions that make a society ripe for noticing and adopting some new technology, though on a different level, are structurally rather similar to the conditions in a single man's mind that precedes his first perception of significance in a novel experience or idea. And both of these are somewhat similar to the conditions that lead to the nucleation of a new structure in a chemical system ripe for a phase change. In any system, if a "thing" is to be perceived at all there must be interaction between it and its neighbors. A change in either the thing or in the environment may be tolerated by the other or it may force its rearrangement. Whether or not change in units of a structure forces change in the whole is a fascinating question, in very essence hierarchical and not to be understood on one level alone. Whether a small region of misfit in a structure is a residue of earlier conditions or an augur of a revolutionary reconstruction cannot be told at a glance but needs a study of history. A simple two-dimensional section of a piece of grained wood or textured stone fails to distinguish between concavity or convexity and says nothing of the connectivity of the features seen. Thus is it appropriate that some papers in this volume deal with individual techniques at specific times while others see things in widest perspective. Both the overviews of Price and Wertime and the specific detail of Alexander are necessary to understand the whole, and so are the varied emphases that comes from the use of artifactual or verbal records.

Schmandt-Besserat's account of the transition from little clay cones, cylinders, spheres, etc. (perhaps originally game tokens?) into tallies to accompany shipments of goods, then into impressions of them in clay envelopes and finally through drawings into symbols as the beginning of writing is about as stirring a tale of man's nature and experience as any that can be found in the poets' works later promulgated by the technique.

Most studies in the past have tended to separate the different classes of materials to allow closer study. Wertime's sweeping overview emphasizes their interrelatedness as part of one pyrotechnology, if not in the work of a single craftsman at least in cultural traditions relating the effects of fire common to all inorganic materials. The discovery of processes (often multivalent) is quite as important as the discovery of the individual properties of materials and the finding of suitable uses for them. The fact that all three aspects have always been involved in any application makes materials so valuable as a record and as a common index to man's activities in all their diversity. Perhaps Wertime exaggerates when he says that glass rather than metal was the measure of Western civilization, but it certainly does represent an advanced salient in knowledge of materials. Reading in today's news of the use of glass fibres in advanced telecommunication makes one marvel both at the continuity and the unexpected diversity of technology.

In the field of materials, one can by selection accentuate almost any property that is desired. Curiosity—almost as exploration in the geographic sense—disclosed most of the general types of material long before any theory guided the search. Mechanisms whose construction involved an ingenuity comparable to the underlying materials that existed in pre-classical times were much slower to develop. Perhaps this was because the aesthetic rewards for play were minimal in the first simple mechanisms, and it was hard if not impossible to anticipate utility. Here, Derek Price's instruments show that theory may have led practice in the beginning, but the real explosion of mechanical ingenuity was delayed and it required the intervention of the artist. The development of mechanism became much easier when methods of representation by drawing were improved to the point where novel combinations could be sketched for trial even more quickly than the making and testing of an alloy in a goldsmith's shop. As Hall remarks, "the inventions of the artist provided the technologist with a powerful new vocabulary for portraying and transmitting information about his subject." Not only did the handsomely printed "Theaters of Machines" give technology a place in respectable libraries, but the artist's methods for the representation of three-dimensional complexity permitted play with mechanisms both for the pure pleasure of discovery and for the testing of ideas in a manner rather like the critical experiment in relation to scientific theory. Hall's essay is important in another way, for it covers the transition from the time when history has to be written from the archaeologist's record based on artifact to the historian's kind of history based mainly on written sources. Does all this also provide a clue to the reason for the rise first of natural philosophy and then the change from "low" to "high" technology?

* * *

More thoughts on the role of art in technology and vice versa were unlocked by the opening at the time of the Conference of exhibitions of Persian carpets and of Persian locks at the Huntington Galleries of the University of Texas.[1] The paper by Williams illustrates both the decoration and the mechanism of the locks and discusses the art-historical aspects of their design.

One of the most memorable events at the conference, unfortunately not reproduced in this volume, was the fine luncheon talk by Henry Miller of the Sargent Lock Company. He used his own experience to illustrate how a working technologist accepts society's aims and matches them with technical possibilities, thus introducing the personal angle that is usually missed by scholars studying the larger issues of technological history. Incomplete though the coverage in this volume inevitably is, its lesson is clear: technology has played its vital role in history because it centrally and essentially unites the social with the individual experience and it gives material form. One cannot understand either man or his institutions in their various forms without a study of technology.

[1] Parviz Tanavoli and John T. Wertime, *Locks from Iran: Pre-Islamic to Twentieth Century*, Washington, D.C.: Smithsonian Institution, 1976.

Introduction: Origins of Technology

by

Derek de Solla Price

Mr. Chairman, Ladies and Gentlemen: It is a great delight and privilege to share with you this celebration of the deep antiquity of something that looks so essentially modern as technology. The scientific and technical revolution that goes on in this world of the atom and space and micro- and mini-computers and calculators is not just a new twig on an old tree of human culture. I shall maintain that it is part of the trunk development itself. Furthermore, it is only our western ethnocentricity that makes it look as if all of this modern high technology is something originated by Western white man. As this symposium helps us reestimate, these technologies that rule us today, first took shape in the Aegean, in the world of Islam, in the Orient, and, indeed, going further back, in the ancient river valley civilizations of the Near East. Perhaps one may even claim that the origins of this modern technological process go back to the neolithic revolution, and I remind you that that fundamental change in technology took place as long before the Bible as we are after it.

It is no easy task to help restore a sense of continuity where there is not one abroad in the modern world. But we need this continuity, I think,—and other peoples of the world need it—as an anchor for the many pressing problems of our civilization. Without it, we seem to drift in a culture that is alien and external, a culture of only the privileged and special group, rather than of all mankind and of all traditions.

Now, a word to my fellow academics: many of us that have had the joy and the misery, whether it be as a professor or as a student, of the Ph.D process, know only too well what we may call the "Ph.D. Thesis Syndrome." Often, a very well-chosen research topic of deep significance erodes away during execution and emerges as a remarkably dull, if correct and infallible thesis. It is not the literary style or the scholarship that lacks; some subtle sense of high strategy has been lost by that redefinition of the problem as you pick a precarious path through that which can be done, that which can yield a satisfactory and closed conclusion. The thesis takes its form from a rather inchoate middle game, and the more or less triumphant ending, and all that spirited strategy of the opening that led you to do it in the first place gets lost.

Something like that, I think, has beset the whole disciplinary complex of the history of technology, together with the economics and sociology and philosophy, and of course the archaeology and anthropology of technology. When we first got the idea of the history of technology, a century or more ago, in the traumata of the industrial revolution, the game was to analyze and explain that transformation that had overtaken a human civilization. There was no precedent from previous times, or of other cultures, but slowly we did what was do-able instead of what we wanted, and, I think, in the rapidly increasing population of sub-fields, we lost the original question. We do not even see a single industrial revolution today. There are mini-revolutions in the Middle Ages and in antiquity—indeed wherever you look in detail—and each particular technology is seen in such a complex of socio-economic conditions, that we have lost a great deal. And then on top of all this, the history of technology has a particular occupational hazard: antiquarianism. It just happens that the documents of the history of technology are not written records, but artifacts, and for scholarly study, we do a lot of our work in the business of translating from an artifact into a literature, and that is why especially those who use it as a cussword call it antiquarian. And often the history stops here—we do not see the "go" of it. My own attitude to historiography is that we are called upon to explain why things happened the way they did, and when they did, and it is in that spirit that I now put to you my question: what is the origin of our modern world with its domination of a high technology, loose-linked with science, and with a particular sort of ingenious complexity?

For better or worse, most of the history of science is conducive to such an attitude—much more than the rest of history—for things tend to happen in science because of the *inner* necessities of science, rather than the will and wishes of society and the needs of man. A historian of science has little difficulty in seeing a strong internal order that has dominated the steady elucidation, step by step, of scientific knowledge since the mid-17th century scientific revolution. At that time, there were generated the new forces of socialization of science within the first special learned societies, along with the rapid proliferation of knowledge in papers and journals, which makes science run faster than books. Perhaps even more important was the Newtonian synthesis of a unified mathematical explanation of astronomy and mechanics that has ever since served as a model of methodology for all of natural science.

Still more significant in that revolution came a new role of scientific instruments. Surprisingly, it is not what the philosopher of science thinks. Instruments were *not* used for testing and experimenting in any form or sense. They were not then, and they very rarely have been since. But, like the telescope that started it all, the instruments revealed new worlds for philosophy. The telescope gave one an artificial revelation; things to philosophize about that no man had seen before; a new nature of the universe to argue about. It did not test theories; it told you things the eye had not seen. Other instruments, like the vacuum pump and the electrical machine, gave you conditions of nature that man had never seen before because nature had never created them.

Going back further, we can trace the path of science, indeed, the reason why it came first, in astronomy. We can trace it right back to Babylonian and Greek astronomy, and it is only recently that we are discovering how the whole process started in Babylon. We had dismissed it, because they did not look like Greek philosophical theories, but suddenly we understand in an age where everybody knows a computer-freak and the way their minds work, as we know how computers work with algorithms—the Babylonians would have loved computers. That was their "theory," and it is, in fact, because they got this off first with a sophisticated algorithmic sense that no other civilization has ever had, that we came to a head start in astronomy. If you think of it, science began with astronomy, and that is not a particularly useful science. If science began where Man wanted useful knowledge, we would surely have started with chemistry and surgery, and left astronomy until much later. No, man starts with astronomy not because it is useful, but because of weird properties of the human mind and of nature, and this part of creation gets itself understood first, whatever our wishes. Clearly it is the nature of science, and not its human utility, that dominates. Science, in fact, was not very useful for anything, not technology, nor any application, until about the end of the 18th century. Any stories you have heard to the contrary are probably fairy tales that are being written daily by the non-professional historians of science.

Having said this now, let us attempt to peel back the historical layers, as I have done with science, but this time for technology. It is a very different story, indeed. Some attributes of technology are of very recent vintage. For example, it is obvious that man used iron thousands of years before there was any understanding of the oxidation and reduction process that requires the presence of carbon. In particular, the direct basing of technology on science begins only with the belated scientific revolution in chemistry at the end of the 18th, the beginning of the 19th century, and with the discovery of current electricity, which at first was a powerful new chemical force, and only after that, a motive power for industry. Incidentally, and characteristically, electricity was discovered not because some man sought something as useful as electricity. Galvani was looking for the nature of life, wondering how the life-force worked, when a static electrical shock made frog muscle bunch up. He knew that an electric shock made you jolt, therefore, electricity was the stuff of life, and he was looking to find out how that worked. And out of that came, incidentally, the discovery of electricity, and it is, at all events, only in the early 19th century that chemistry and electrical science enter industry as an armory of new materials and new forces for people to do things with.

In the stage before that, you do have the methods of the scientific revolution applied to industry in order to understand and improve the crafts using the new methodology of scientific instruments and inquiry. That is the sort of thing that began with Bacon and flourished in the enlightenment. It was not the promotion of pure science to applied; it was the promotion of research methods to the technological crafts themselves, and it is in that spirit, for example, that we get people examining the first Newcomen steam engines, and finding out what parameters of size and temperatures would give you the most useful, the most efficient steam engines for industry. They did not have the science to apply, they did research on the technology, as they did on the production of steel, without understanding fully the chemistry involved. Before that, again, peeling back those layers, we still have the germs of modern high technological civilization. The scientific instruments of the scientific revolution came from somewhere along with the ingenious crafts. Where did all this come from?

The answer, I think, lies in a very ancient tradition of instruments and ingenious mechanical devices that were on the whole utterly useless except they gave us ultimately the computer, the atom bomb, and the space age. Curiously, these early instruments and devices were not at all utilitarian. It is a mistake to see any early instruments as measuring devices and tools. They exist, as I shall now illustrate, as a sort of philosophical technology, a simulation of nature, which philosophically is a tangible version of a theory. Now ask, "What is a theory?" You say, "It is some sort of explanation." Fine. How do we explain? We explain in terms of a model, be it a mathematical model or a mechanical model. And what I am now going to say is that deep in antiquity, man used and developed special crafts to make models—craft models, tangible things—which for them constituted explanation, participation, a sort of do-it-yourself creator kit, of which an Einstein or a Newton used a mathematical equivalent, a model that gives you participation in creation.

This is the point at which I say the techniques of high technology had their origin. It is the sort of technology that is not a steadily-improving utilitarian craft, but a technology that embodies science, and it is this that is at the roots of the ingenuity of the mathematical practitioners of the 17th century and at the roots of the Yankee ingenuity. It is the ingenuity of countless inventors who made better mousetraps, and the surprise is that this whole cultural tradition is very important and makes our civilization in many ways different from most of the others because of the ultimate dominance of this force. This is a technology-based science that we are dealing with, rather than the science-based technology, which is but a Johnny-come-lately accretion to the main cultural line. In the beginning it must be realized that technological crafts were already ancient when writing was invented. Thus, by definition, most of modern technology is pre-historic in its origin since the culture of science is equally clearly *post* the invention of writing and a written cultural heritage. It is quite clear that technology takes a very long chronological precedence over science.

The first powerful science was this heavily mathematical astronomy, incredibly sophisticated, of the Old Babylonian civilization of the fertile crescent. This had theory—which as I said, we only now begin to realize—a computer, algorithmic type of modelling, a purely intellectual argument. With the other river-valley civilizations of antiquity, they did not have this other-hemispheric-dominance attitude that makes computer whiz kids so different from the rest of us. They thought in pictures and they understood, as most of us do, by seeing it, not by hearing it, and following the sequence of equations. So it is with the other civilizations that we know a lot of, where we first meet the strain that I wish to report to you.

In Egypt, in the Aegean, and in Greece, we get for the first time, these philosophic crafts. With no sophisticated mathematical algorithms, the first thing we get is a modelling of astronomy. Perhaps it goes back very much earlier, but from Egypt, from the Old Middle Kingdoms, we get star ceilings, and depictions of the heavens. The Dendara Zodiac is just static, just a picture; perhaps it is not much more culturally than the much more ancient cave paintings, whose significance we do not understand, but which I now believe as being more scientific theory than a religious good-luck wish in hunting. The star ceiling is an attempt, I think, to take something which we cannot see, the whole heavens, we see them only rising and setting, and it depicts the whole of creation in a way more than man can see at one time. The first mechanization of the star ceilings, a clepsydra, a leaky water-pot, that used to be thought measured time. It does measure time if you look at it from a modern point of view. From the ancient point of view, though, I think this was a *theory*. It was a theory that the heavens revolved with the same steadiness that water runs out of this very sophisticatedly designed pot; whose flower-pot shape makes it go like uniformly moving heavens.

Illustration 1 shows the inside of the sun-dial pot with its water clock part with its graduations that made it look as if it were a measuring instrument. When you look at them, they are all astronomically oriented rather than being a useful time measure.

I cannot show you the earliest Greek instruments because these were very fragile, but the best I can do is an example from the 15th century that gives us a replica of what we know was there in the 5th century B.C. Perhaps it is the world's first teaching instrument (plate II, ill. 2). Unfortunately, astronomy is three-dimensional and people are very bad at seeing three dimensions, so you must make them a model of the standard circles of astronomy, an armillary sphere. This the Greeks had from the beginning, and again, it is beyond being a teaching aid; it is perhaps a theory. When you have a model like this you understand how the circles work and how the heavens turn.

The sundials also only mistakenly looked useful. Sundials appear in Greece from about the 4th century B.C. onwards. One example (plate II, ill. 3) hangs on the edge of the Acropolis and looks as if you could use it to tell the time for going to the theater. Actually, it is again a monument for creation, because as Sharon Gibbs, who studied these dials, discovered, they do not tell you the hours. What they do, when they start lettering them, is to give you all the astronomical circles and the signs of the Zodiac, and the purpose of them is to show the annual cycle of the sun, reflecting by its shadow, the great circles of the astronomical mathematics by which we understand the very orderliness of that creation. One could see the sun in its cycles and that assured you of the mechanicalness of the astronomical creation.

These sundials become extremely complicated and we have something of a hunch that making them to give a proper reflection of the heavens is what really started man off inventing the geometry of conic sections, because it turned out that the cone was a much prettier thing to work on than the sphere. You can draw straight lines on a cone and you can't on a sphere, and you can draw straight lines on paper but you can't draw all these circles without making a very complicated geometrical construction. That, then, is how it all started with astronomical models.

The next step was the continuance of this tradition to the world of Islam. First, we have the spherical astrolabe, a model of the spherical heavens on which you can not only see what happens, but also make the measurements and calculations and see it agree with the heavens, not just qualitatively, but quantitatively, too.

This then becomes what looks like an astronomical globe. We know what a globe is for, but we don't realize that the original purpose of the globe is still very much like the purpose that brings a person to use a globe as a decorator object—not to find one's way around, but to encompass the heavens or the Earth in your own living room; to know that you have got it all in your "possession." The Islamic globe is, I believe, originally a little device that was held in a handle. A little knob that was on the top was, I think, originally on the bottom. Anyhow, it becomes a handle and goes into the next object, which it turns into a flattened globe, or astrolabe. The earliest-known astrolabe (plate III, ill.4), is a flattened globe, and, strangely, it is through this very instrument that we have been able to determine that the whole tradition goes back to early Greek days, instead of starting in Islam. The astrolabe itself had Greek and Roman counterparts.

We have a little fragment that was part of a monumental eight-foot diameter stardisk from Salzburg that belonged to a water-clock in late Roman times (plate IV, ill. 5), and this enables us to reconstruct one of the most curious and beautiful objects of the Greek world, the Tower of Winds in the Roman Agora in Athens (plate V, ill. 6).

This is one of the key objects in this whole history. The reconstruction shows that it was built upon an octagon plan with a little circular water tower at the back, and its shape is not accidental. The octagonal form is to *prove* the theory of creation. The theory of creation has to do with the element scheme—which we have from Aristotle, but which certainly goes back earlier—of the two fundamental forces of the Universe working at cross-purposes, hot and cold, wet and dry, giving you Air, Earth, Fire and Water. When you really understand it, you know you could have started the other way around with Air, Earth, Fire, and Water, the fundamental forces, and produced the hot and cold, wet and dry forces, which, by the way, in the form of the *cross*-purposes is probably the origin of the mystic significance of the Christian cross. In this later form, the cross becomes a sort of double-cross of a square diagonally within a square, and because of that, the Greeks knew that there *must* be at least eight winds, that there must be an octagonal symmetry in the Universe, and therefore, this was the shape, so to speak, of the fundamental equations. This was the Greek equivalent of the Schroedinger wave equation, or the basic equation of relativity. The Universe was like that, and the Tower of Winds was built as a monument of it. Inside it has a sort of Salzburg Disk, or water-clock, working an astrolabe that turns all by itself, showing the heavens rotating behind a grid; displaying the sun even at nighttime, and the moon and the stars by day; showing you all of creation, the Earth and the Sky or Air worked by a water-clock. I always wondered why the Tower of Winds had no windows, but then it occurred to me it had to be dark inside because they had to use fire to illuminate it, as the fourth of the elements. Thanks to a Turkish traveller, we know that the whole thing inside was full of the symbolism that enabled one to understand the Greek knowledge which was practically perfect (they thought perfection was around the corner) of a complete scheme of the Universe in its physical aspects. Now that achievement, coming from the modelling of the Universe, continues.

We have almost a copy, it could be, of the interior clock of the Tower of Winds, but built in Fez, Morocco, in the 14th century, in a tradition that we now know must have been transmitted, perhaps through manuscript descriptions. Because the technical details so faithfully preserved the much earlier ones that we have now recovered, we can see that there must be a continuity of master apprentice tradition of this particular sort of craftsmanship. There is an astrolabic dial, like the dial within the Tower of Winds. There is also a row of windows out of which pop little mannequins who beat gongs and blow trumpets. Those are worked by a long wooden channel. We can see in the drawing of an instrument like this by al-Jazari who lived in Haran,—now in modern south Turkey, just inland northeast from Aleppo—and seems to be the point at which we can pinpoint a synthesis of Babylonian, Greek, Byzantine, Persian, and Arab influences in a new technology of this sort. The same technology thus produces a clock face with all of these trumpets in the same row of windows and funny little birds that sing and these are all linked up to the birds and devices on the clock face. One wonders what this tradition is doing. All of these little birds worked by water wheels and had whistles worked by pneumatic energy produced by the water blowing the air up in the canister.

The whole idea goes back to an illustration from Haran and much farther back in history, I would claim, even to Old Kingdom Egypt where they were having a do-it-yourself attempt upon the other half of creation. You make living beings, especially little birds, for some reason, because they look like the elementary "talkers" of the time and you can make them flap their wings and you can make them whistle and then you can have all this display of the elements. That, then, is a main story that goes all together.

To jump ahead a little, let me encapsulate it by saying that is why there is a cuckoo on the cuckoo clock. The cuckoo is a direct descendent of this bird and the little aperture that gives you the phase of the moon is another part of the astronomical showpiece. It is not for useful purpose; it is not even for fun; it is again the Schroedinger wave equation of its time.

One of the most spectacular and wonderful objects from antiquity is the Antikythera mechanism (plate VI, ills. 7a,b), a strange device of huge clockwork complexity which came out of the very first bit of underwater archaeology in 1900 when sponge fishers discovered one of the richest treasure ships ever found, from which came a great many of the finest pieces in the National Museum of Archaeology in Athens. This little object had been a mystery for years with its astronomical inscriptions in perfectly good first century B.C. Greek and its very clear wheels and gearwheels. What on earth could it be? It presents a weird mess of gearing within gearing, and for something like twenty years I tried to put it together. Then I was lucky. We got some x-ray pictures and those showed that there were many more gear wheels inside where you could not see them and the pictures were so good that we could count the gear teeth and put it all together. Altogether, it is the most hopeless mess of gearing that you have ever seen. It is as complicated as anything in a piece of 18th century industrial revolution machinery and clearly that is where the clockwork part of that revolution came from. It turns out nobody could design it in such a mess. If you look at it the other way around, in plan, it is all very rational and you find out that what it is producing, the hard way around, is a duplicate of the way God created the cycles of the sun and the moon. We know that for every nineteen revolutions of the sun, there are 254 revolutions of the moon, and that produces 254 minus 19, that is, 235 alternations of the phases of the moon. If you just wanted to calculate, you would use gear wheels with 19 teeth, and 254 teeth, and 235 teeth and you would get a perfect model. Such models turn out to be a very long tradition that begins at least with Archimedes—and I suspect really begins there, and was elaborated in fits and starts—and ended up being a craft tradition in first century B.C. Rhodes. The people who worked on this did not want to do it to make calculations; they did not want to do it just to astound. They wanted to do it to duplicate the way it was done in the heavens. So they made the 19 revolutions, geared it to make the 254 lunar rotations, and then, built into that device, they used a differential gear—now that is a very complicated device—a differential gear to give you the difference between the sun and the moon and then they put everything on dials so you can see it. The machine had two dials in the back and one in the front and all the clockwork in the middle. It was about the size of a portable typewriter, but it was obviously designed to be mounted up on a stand as an art object. I strongly suspect that it was held in the arms of some Atlas or other statue that was in the shipwreck and I strongly suspect that it was part of the cargo that young Cicero was taking away from Rhodes when apparently he lost his baggage on the way back to Rome.

Where does all this go? The differential gear that I take out there as part of the mechanism—a complicated device—is perhaps the first mechanical sophistication that is completely non-obvious. You have to think a lot and play with gears a lot and want to do it like the heavens do it rather than usefully to invent that.

This idea turns up in the whole tradition of clockmaking. Let us look at that tradition when it comes out of Islam. You get people like Richard, Abbot of Wallingford, a blacksmith, a craftsman, a person who was an heir to this tradition of doing complicated devices, who spent his years as Abbot—and apparently most of the money of the Abbey—on making what is called a clock, but didn't tell the time. It showed all the planets' movements.

A similar idea a little later is found in this clock of Giovanni de Dondi, 1364, who made it in Padua. A reconstruction can be seen working in the Smithsonian Institution. The tradition went on to mighty clocks with bells like that, that again were more *theories* than time-tellers. This is from a *Clock of Wisdom* manuscript of the late Middle Ages. And the tradition then erupts in the great clock of Strassburg. This is why those great clocks were in the cathedrals: not to time the sermon, but as an altar-piece to the scientific creativity of man. One must remember that it was only after Galileo lost his fight that the Church failed to become the chief promoter of science. If only Galileo had won, we would have as an outcome of these cathedral clocks that all physics labs and computers and telescopes and atomic power stations would have a crucifix over the front door.

Another of the famous clocks is the one in Venice. And it goes on from these monumental ones to the great clock at Wells that looks much more workmanlike and is obviously the ancestor of the machinery of the Industrial Revolution. And the line goes on from that to one of the last of the pre-transistor clocks that told absolutely everything about the date of Easter and the conjunctions of the planets and there was even a little dial around the back that told you what time it was.

The other line, the automaton, the simulation of living creatures, also continued as a history. We get in the manuscript of Vilars de Honnacourt these little automaton devices that move by themselves. The line results in this little automaton lady, one of the earliest of the really complicated automatons that do a lot of things. This is about 1550.

Those automata are very important not only because they were the point of origin of a lot of machinery—I mean, for example, Vaucanson invented rubber tubing in order to simulate the gut of a model bird—a model duck—that he was making. And where would chemical labs be without rubber tubing?—and not only for that sophisticated technology part but also because it provided the idea that living man was an automaton. It produced the deterministic theories of biology. We go from all of this tradition to the people who made these automata, these clocks, these sundials, and astrolabes. Here is one long before Francis Bacon—Gemma Frisius—in 1555, about contemporary with the automaton there. These are the sort of instruments that he made.

The idea, the simulation of nature, is the same as it was in the earliest examples we have, but now the dials have been host of all of the new craftmanships that have developed through the Middle Ages and the Renaissance—the ivory work, paper work, the metal work, the engraving, the woodwork, and the decoration with gems. They are all pocket pieces, pretty objects, and almost religious rather than useful instruments. They are tangible theories and although by now they also begin to be useful, they have an attractiveness, a fascination which is particularly evident as a motive spirit of that Industrial Revolution. It is the same craftsmen, after all, who produce this ornamental lathe which for its 16th century is just about halfway between the exuberant *joie d'esprit* of artisan crafts in the service of scientific theory and the draftmanship knowledge which through the use of machine tools like this is going to produce all of the instruments of the Scientific Revolution, first, and then the Industrial Revolution.

To close, let me just give one very brief example of just how it is that this line of tangible theories of technology in the service of science produces the very things that are the most central in the Industrial Revolution that happened as the end result of this line—or not quite the end result—the post-World War II high-scientific-technological revolution is just the next stage. The Antikythera machine with its differential gears is one of the few pieces of technology we can fix on and say exactly who it was who carried it from there to there. As a differential gear in astronomical clockwork, it survived in all of the most complicated clocks of the Middle Ages and was particularly well-known in the Renaissance except that in the Renaissance people began to have trade secrets and therefore it was rediscovered again and again—what the anthropologists call a stimulus diffusion because we don't really know how it happened.

However, we can take the story from the clockwork. Asa Arnold and a couple of Englishmen who were trained in the clockmaking tradition, who knew this device as part of the armory of a hundred-and-one bright mechanisms for the inventor—they had this in their toolbox, so to speak—and when the textile revolution was just about to erupt and people were taking the craft of the wheelright and applying it to the very difficult problem of mechanizing what had been one of the most ancient of the craft technologies, they realized then that the peculiar properties of cotton needed gearwork with a sort of elasticity that wouldn't break the thread when you spun it, and in the bobbin and fly frame it was crucial to use a differential gear to take up the slack. It was the use of the differential gear that made possible the mechanization of the thread industry and gave rise to the invention of factories, the invention of the working class, and all those other changes that happened in late 18th century England. It was then a well-known and rather mysterious technical device, a technical secret of the textile machine industry, the sort of thing that industrial espionage like that of Eli Whitney could take from country to country.

But it was Richard Roberts, who had grown up as a textile machine craftsman, who got involved with the early steam engine and started working on steam automobiles, who realized that you needed in the automobile something not only to make the wheels drive, but turn corners, and that very thing he needed was his textile machine differential gear. And that is how they got into the back axle of every Detroit motor car. The differential gear that goes all the way back to Greece and was motivated not as anything that was useful, but as a philosophic model theory.

I rest my case there as a historical elucidation of a rather peculiar and unexpected line. That, of course, does not tell us about the steady evolution of the low-technology crafts. That's quite a different story. It is one that precedes science and is not affected by science until very modern times. The tale I have told is of course only part of it. It is overlaid by all of the new scientific things. But the reason why our civilization left the path of all the others is the presence in ours alone of this particular main line. It happened also in China but with that civilization they didn't have a Ptolemy, not an ancestry of Babylonian astronomy, and therefore, they could not have their Newton. And anyhow, in that civilization, as in most of the ones we study, their Hitler won and civilization was interrupted. In ours, we have the continuous story, and I beg leave now to rest the case that our high technology is something that was essentially developed by the Middle Eastern river valley civilizations in the ancient Aegean. We have never known it before, perhaps because we have always looked back with the unhistorical utilitarian eyes that are too indoctrinated with modern science.

I. ART AND CRAFT OF MATERIALS

Pyrotechnology: Man's Fire-Using Crafts

by

Theodore A. Wertime

Four materials defined man's early fire-using crafts. In order of appearance they were baked clays (9000 B.C.), plasters (9000 B.C.), smelted metals (6000 B.C.), and glazes (5000 B.C.).[1] These dates are necessarily approximate, round figures derived by radiocarbon at a number of sites in the Middle East. Only in the old Fertile Crescent and the Eastern Mediterranean do we have all four technologies reaching a climax in some necessary technical juxtaposition (for all are interconnected by the chemistry and physics of oxidation, reduction, heat aggregation, and heat separation of materials). Only in this area did iron and glass and cement appear with some simultaneity (1400-1100 B.C.) as the culminating moment in the evolution of what may be called fire-using industries (*industries à feu*).

In Southeast and East Asia there are claims for the early development of both ceramics and bronze metallurgy. But only in Southwest Asia and Egypt did baked clay play a direct role in writing and architectural systems, and metals gain a paramount influence in the origin of trade, banking, credit, money, and economic life in general. Only in the ancient Mediterranean under Greco-Roman control did civilization climax fully in the simultaneous employment of iron, glass, ceramics, and plaster-cement. Only in the European arena did such industries grow toward their modern manifestations ranging from Venetian glass to the steel of the Industrial Revolution. I include China in this comparison.

The terminology in common use for this period of western civilization in the Mediterranean and Europe is "Iron Age," as the legitimate sequel of Bronze Age and Stone Age. But "Iron Age" obviously is a misnomer for the magnificent procession of objects of glass that came to mark ancient Rome—tableware, optical glass and even window glass. And again "Iron Age" fails to capture the tenacity of the cement incorporated into the Roman Colosseum. Nor does the term tell us much about the long ups and downs of steeling of iron, which was identified by 1000 B.C. but which did not become a mass-produced art till the mid 1860's A.D. when Henry Bessemer and William Siemens perfected their respective furnaces.

From the standpoint of energy and materials, we may better identify this period of emerging western civilization as the Age of Pyrotechnology: the age of the employment through fire of photosynthetic fuels for the oxidation, reduction, separation, and aggregation of earthy materials. Seen in this light, the history of material culture in western civilization can better be interpreted as the only twice interrupted geometric ascent in the *industries à feu*. They were first interrupted by the Dark Ages in the Eastern Mediterranean after 1000 B.C. and again in the Mediterranean and Europe after the first sack of Rome in A.D. 410.

One can also gain a better perspective on the three crises in energy that have marked the history of the uses of fire in the west: first the depletion of wood along the Mediterranean littoral and in the

[1] Based on studies of the author. The story of the consequent deforestation of Europe is in Wertime. 1962. *The Coming of the Age of Steel*. Chicago: University of Chicago Press, pp. 113ff.

mountains and deserts of the Sahara and Middle East; second the 40 percent deforestation of Europe and parts of North America between 1400 A.D. and 1800 A.D.: and finally the present imminent exhaustion of the world's supplies of fossil fuels. In this perspective, the Age of Pyrotechnology has about run its course in exploiting man's course in exploiting man's dependence on processes of oxidation and reduction.

A final phenomenon of the Age of Pyrotechnology has been the thorough demolition of the so-called angel theory of human technology. More will be said in the closing section on the forms of technology that have dominated human history: stress technology (viatechnology), water-based technology (hydrotechnology), chemotechnology, and the like. The evidence now pouring in from dozens of sources, such as Barbara Purdy at the University of Florida and Denise Schmandt-Besserat at the University of Texas, makes clear that Stone Age men were using fire in manifold versatile ways from 25,000 years ago. These ways ranged from the tempering of wood spear points, the oxidation of such pigments as ochre, the annealing of stone projectile points, the fire-setting of quarries (from which the lime kiln and possibly the metallurgic furnace emerged), to the development of the technologies of the cooking hearth. When the full-scale baking of earths for ceramic, bonding, vitric, or metallurgic purposes began some 8,000 to 10,000 years ago, artisans were already launched full-scale into the chemistry and physics of materials.

If they had been previously familiar with the temperatures of annealing of materials to reshape their crystals, or oxidation to reshape their colors, they came to learn temperatures of melting, forms of reduction of oxides, chemical combinations of elements (iron oxide with silicon, iron with sulphur, etc.), the electropotential of certain elements, as well as the complex relationships of carbon with iron or of lime and soda with clays.

One can no longer see pyrotechnology as the movement from the simple to the complex, but as the separation of single strands of new arts and crafts (such as glazes), from complex skeins of existent fire-evolved chemistries and physics.

Metallurgy is often viewed as a craft moving in linear fashion from the hammering of native copper to the melting of the metal and smelting of the ore.[2] In fact, the first annealing of copper brought craftsmen into a long extant tradition of fire-using that at once enlarged its horizons enormously. But then again native copper may in many cases have been broken from its matrix by fire-setting, already making clear the potentials of annealing and casting (as well as the reduction of ore in a furnace or hearth).

Iron was certainly the centerpiece of this era, even if one cannot capture all its ramifications under the heading of Iron Age. From perhaps a fraction of a ton a year in Cyprus, Greece, and Palestine in the year 1000 B.C., 50,000 tons a year were being yielded by A.D. 1, mainly in Rome and Han China and Sassanian Persia and Hindu India. By A.D. 1700 the world was producing 300,000 tons of iron and steel, by 1800, 500,000 tons. To show how geometric was further growth in the Industrial Revolution, the figure was 352 million tons in 1953, and 1 billion tons in 1970. Today iron accounts for about 5 percent of the world's energy consumption.

In the year 1700, when wood was still the dominant fuel, iron may have accounted for 25 percent of world uses of energy. So large did it bulk in the total picture of energy during the eighteenth century that the industry helped to push England toward the enclosure movement and the Industrial Revolution and France towards the French Revolution.

Wood was the fuel par excellence for the revolution in the uses of fire in shaping earth. Dung had played its part in the hearth in accomplishing the miracles associated with the baking of a wheaten bread. But the native stands of wood in the semi-arid Mediterranean littoral fueled the fires of the new technologies. The thermal values of the wild pistachio of Turkey, the *acacia arabica* of Egypt, the wild

[2] Some persons have interpreted my 1973 article in *Science* on the beginnings of metallurgy as an argument for a simple linear sequence of technical steps, when in fact I was arguing the complexity and interrelatedness of early pyrotechnologic metallurgy.

almond and *haloxyon amodendron* of Iran, the oak of the Greek island of Kea, and the Mediterranean pine of the Balearic islands, Elba, and Laurion encouraged sequential development of materials. In time glass and bronze became the exhibit materials of Middle Kingdom Egypt, Elban iron the mainstay of Etruria and early Rome, and silver the economic crutch of Periclean Greece.

One can find the main smelting furnaces for the lead of Laurion near the ancient wharves where the wood or charcoal was brought by ship from northern Greece. The story of erosion in the Mediterranean now being investigated by Colin Renfrew on the island of Melos tells us of deforestation as early as the Iron Age (around 1000 B.C.).[3] Plato had his views of the defoliation of Greece. In Egypt the wadis of the Eastern Desert with their occasional acacia tree bespeak the long despoliation associated with the pyrotechnologic industries. Perhaps 500,000 tons of slag near desert Nakhlak in Iran tell us of a similar fate of the once green (if semi-arid) biomass of the now salt desert.

One can follow the decline of Laurion silver for Greece or Elban iron for Rome as a phenomenon of diminishing fuel rather than of ores. The goat has long been blamed for the faults of the pyrotechnologic furnace; he only consolidated the damage.

The net effect was that, outside of such unavoidably rich sites as Rio Tinto in Spain, Rome consolidated its metallurgy largely on the forested flanks of Europe. That Rome drew lead from England or copper from Illyria or iron from the Slupia Nova mountains of Poland was as much a matter of fuel as it was the skills with metals of the Indo-European tribes, such as the Celts and Germans. For European metals were as much a means of cutting the forests of Europe, tilling its river bottoms with new forms of plows, and harvesting the wheat with wheeled harvester, as they were their inspiration.[4]

Oliver Davies[5] finds a major symptom of the decline of Rome in its loss of control of management of these mines: setting the stage for the European tribes to develop their own post-medieval industrial revolution and consummate the deforestation of Europe by A.D. 1800. But Late Rome was on the edge of at least one great breakthrough in materials, the general introduction of window glass into Europe.[6]

Until recently many schema have been presented for the appearance of pyrotechnology in the urban revolution, with the general observation that the ceramic kiln seemed to have priority in the organized heating of earthy matters. In 1942 H. H. Coghlan concluded that the first accidental smelting of copper took place in the pottery furnace, giving rise to a generation of speculation that metals took their cue from ceramics.[7] More recently Gourdin and Kingery of the Massachusetts Institute of Technology have speculated that the lime kiln preceded the pottery kiln and helped to fashion it.[8]

In fact the varied stone-age avenues to pyrotechnology mentioned at the beginning each sparked experimentation and interplay between forms of the use of fire: kilning was only one of several pyrotechnologic arts being tried in the late Neolithic furnaces found by excavation. One can see Maden Cörüfü in the museum at Alaça Hüyük in Turkey, a primitive copper smeltery surrounded by buns of copper and fayalite slag and dating to 1800 B.C. with antecedents in Anatolian history. He can survey the various furnaces at Timna in Israel going back to the coming of the Midianites and the beginning of large-scale copper metallurgy. He can speculate with Coghlan that men threw chunks of colored ore into a hearth fire and in time came upon the trick of reducing copper or lead, especially if they added an

[3] Donald Davidson, Colin Renfrew, and Catriona Tasker. 1976. "Erosion and Prehistory in Melos: A Preliminary Note." *Journal of Archaeological Science*. 3:219-221. J. Donald Hughes. 1975. *Ecology of Ancient Civilizations*. Albuquerque: University of New Mexico Press, pp. 68ff.

[4] For example, Ralph M. Rowlett. 1968. "The Iron Age North of the Alps." *Science* 161:123-124. Georges Duby. 1974. *The Growth of the European Economy*. Ithaca: Cornell University Press, p. 15.

[5] 1935. *Roman Mines in Europe*. Oxford: Clarendon Press, p. 4ff.

[6] Alexander Nesbitt. 1879. *Glass*. New York: Scribner and Welford, p. 18.

[7] H. H. Coghlan. 1956. *Notes on Prehistoric and Early Iron in The Old World,* Oxford; University Press, p.45.

[8] W.H. Gourdin and W.D. Kingery. 1975. "The Beginnings of Pyrotechnology: Neolithic and Egyptian Lime Plaster." *Journal of Field Archaeology*, 2:132-150. Even as Gourdin and Kingery were preparing their article on kilning of plaster, they were anticipated by John S. Kopper and Guillermo Rossello-Bordoy. 1974. "Megalithic Quarrying Technologies and Limestone Technology in Eastern Spain," *JFA* 1:161-170.

automatic fluxing agent like limestone or iron ore. Such speculation would not be entirely wrong, nor would it be entirely right. For, as noted previously, men had so long known of fire-setting of stone and native copper, been so long aware of the annealing of materials, even encountering fluxes, that the idea of *de novo* experimentation with metallurgic ores around 6000 B.C. becomes slightly ludicrous.

Again with respect to glazes and glass, the sudden eruption of cored glass vessels in both Mesopotamia and Egypt toward 1400 B.C. has aroused curiosity as to how a technology could, as it were, spring like Minerva full-blown from the brow of Jupiter. The issue is discussed at length in the volume by A. Leo Oppenheim. Robert Brill, and others on the glass texts of Mesopotamia.[9] Pliny the Younger, Roman naturalist, offered one speculation that is not without merit: that traders camping on the natron-rich sands of the Lebanon with their fires accidentally vitrified the sand (Strabo says the same).[10] Robert Brill has supplied a number of thoughts about the evolution of glazes into faïence and Egyptian blue. The mystery remains that practically no glass appears at archaeological sites from the fifth to the second millennium B.C.[11]

Today we accept that the smelting of metals with alkaline fluxes and the slagging of those self-same metals created a variety of possibilities for glazing, as did the kilning of pottery in juxtaposition with lime or lead. That the first Egyptian cored vessels had a lead-antimony colorant bespeaks a long association between metallurgy and the vitric arts. There was no such thing as a completely "sanitized" furnace for metals or kiln for pottery or glass, though chemistries varied enormously from one material to another.

If the old Copper Culture Indians of North America were setting fire under taconites and native copper in the Minnesota-Michigan Peninsula, one can have no surprise that Hopewell Culture and its successors present us with fire-setting of everything from jaspers to copper over a terrain extending from Alabama to the upper Midwest. The cutoff for the North American Indian was the upper annealing temperature of copper, namely about 1000° C. He never quite made it to the crucial melting point of copper, 1080° C, nor to the crucial concept of a furnace or kiln.[12]

This seems to have been that turning point at which men's minds were caught by the possibilities of melting metals in crucibles, smelting and slagging ores in the hearth, playing with frits and glazes, experimenting with high thermal fuels, and gaining an acquaintance with iron ore and lime and soda as three important fluxes in separating silica from the matrix of metallurgic ores.

How, then, can one reconstruct the beginnings of the process of using fire that towards 1400 B.C. led men to those enormously complex and virtuosic products, glass and steel? Today, the process must be seen as going back to man's very origins. In 1948, at Kariandusi in the African Rift Valley, Louis Leakey made the discovery of a number of hand axes and cores of the volcanic glass obsidian.[13] These are now dated by magnetic shift to between 700,000 and 1 million years of age. In the years intervening since 1948 little thought was given to obsidian as a material that might in time have set some of the standards for the incipient Age of Pyrotechnology.

Both archeology and the study of materials, however, have converged on nature's glass as an important and continuing pace-setter for the old Stone Age, one possibly as important as the slowly evolving cooking hearth. Through the burgeoning archeology of the Rift Valley of Africa, particularly in Ethiopia, it is now recognized that obsidian has been in continuous and widespread use since the Acheulian, and possibly since the evolution of early man. Ethiopian villagers still avail themselves of the material for

[9] A.Leo Oppenheim,et.al 1971 *Glass and Glass Making in Ancient Mesopotamia*. Corning: Corning Museum of Glass, pp.4ff., 76ff., 105ff., and 181ff.

[10] Gaius Plinius Secundus. *Natural History*. Translated by D. E. Eichholz. 1962. Cambridge: Harvard University. Book 36, vol. 10, 157-158. H. L. Jones, editor. 1961. *The Geography of Strabo*. Cambridge: Harvard University. Book 16, vol. 7, 271.

[11] See Oppenheim, footnote 1.

[12] D. L. Schroeder and K. C. Ruhl in *American Antiquity* in 1968 (33:162) analysed Old Copper Culture and Hopewell artifacts and found only annealing temperatures.

[13] Data on the Kariandusi hand axes and their dating was provided the author by John A. J. Gowlett, who has collaborated with W. W. Bishop in their study.

cutters and hide scrapers, as being superior to scarce metal. And superior it is, according to recent University of California studies.

In the emerging picture of obsidian in the Old World and New, there are reasons therefore for accepting the prominence given by Romans to Ethiopian obsidian, even the name, which according to Pliny is a corruption. Ethiopian obsidian was so prized in Rome that life-sized statues of Caesar Augustus were carved of it and a tableware of artificial glass developed in imitation of it.[14] We still do not know how the long history of obsidian in Africa may have conditioned the development of obsidian in the Mediterranean, as now recorded by Colin Renfrew and others. The natural pathway would have been the Nile; and there are ample evidences of the use of obsidian in Egypt or the Sudan.

Be that as it may, it is clear that from the Solutrian period in Europe and PaleoIndian period in America, other stony materials were being tempered by fire to give them the flaking and chipping qualities of obsidian. Glass when struck by a hard instrument tends to shatter vertically. But when gently knapped by wood or bone, glass flakes laterally, and evenly, providing the experience noted frequently by such knappers as Don Crabtree and Bruce Bradley. The effort to duplicate this condition in flints, jaspers, cherts, chalcedonies, and quartzes seems to have afforded the incentive for fire-annealing of stone that is now so widely documented in Europe and among American Indians and which supplied many of the antecedents for the heat treatment of clays.

As Barbara Purdy of the University of Florida describes it in her various papers,[15] Florida cherts were brought to a temperature of 350-400 centigrade to drive off the water and consolidate the crystals, often but not always yielding a slightly vitreous color in the stone and evidences of the reddening of iron oxides. These are preparations for pottery. Clearly, men were *not* on the edge of making glass (though glazes on stone are a common phenomen of buildings burned in large conflagrations, such as Celtic forts in Europe). They were seeking, with success, to approximate roughly the physical performance of volcanic glass under conditions of stress. For the heated cherts do flake with greater predictability.

If volcanic glass was the first pyrotechnologic material to have a pervasive effect upon the Stone Age, artificial glass was the last of the four man-made pyrotechnologic products to find a niche in an increasingly metallurgic civilization. I tend to discount the absence of glass in archaeological sites as being analogous to the absence of early iron, though Jane Waldbaum has now made an impressive collection of archaeological evidences of iron, going back to the fifth millennium, for the *Coming of the Age of Iron.*[16] Both iron and glass were recent materials, iron gaining an edge in the frontier of religious totemism because of the burgeoning interest in metals.

The advent of cored glass was almost entirely an esthetic choice, a gamble of artisans in Babylonia and Egypt that the sumptuary interests of a kingly society would justify a fragile imitation bronze vessel. One can say that with Roman tableware, glass could be perceived as less of a luxury than before. But not till the advent of window glass did glass become a real necessity. Window glass was used extensively in the Roman houses along Hadrian's wall in Northern England.

The appearance of metal meant a gigantic change in the psychology of man's use of materials, and the utter abandonment of the obsidian standard. Native copper begins to deform plastically the first moment it is struck by a hammer, and without any indication of shattering, whether vertically or horizontally. In one fell swoop, the first adventurers with copper (or other native metals) violated the basic canons of the Stone Age. Indeed they introduced a material that was far less utilitarian than is commonly supposed. Copper axes are better than stone axes for cutting trees. The experience of the Ethiopian villagers teaches, however, that obsidian razors and scrapers are sharper and hold a better edge than does copper.

[14] Book 36, vol. 10, 149ff.

[15] See 1974. "Investigations Concerning the Thermal Alternation of Silica Materials." *Tebiwa*, 17:37-65. Also Don E. Crabtree and B. Robert Butler. 1964. "Notes on Experiments in Flint Knapping: Heat Treatment of Silica Materials," *Tebiwa*, 7:1-6. J. Jeffrey Flenniken and Ervan G. Garrison. 1975. "Thermally Altered Novaculite and Stone Tool Manufacturing Techniques," *Journal of Field Archeology*. 2:125-131.

[16] To be published by Yale University Press.

The advent of copper and bronze therefore must be seen as a gross and silent slippage into new ways, brought on by esthetics as well as utilitarianism. Old Copper Culture Indians may have slipped naturally from the fire setting of the taconites of Minnesota (a very contemporary material) to the fire setting of the native coppers of Michigan. In a recent masterly chapter for the forthcoming *Coming of the Age of Iron,* James Charles has given us a color chart of the metallurgic minerals and of the order in which men, like flies in amber, might have been immersed in them. For example, similarities in color might have led men from the arsenopyrites and arsenical coppers to the tin-bearing copper, stannite. But there was much more to it even than that.

For bronze, we are now discovering, was very much an elitist metal, useful to the ruling classes for war and ornament, and little applied to agriculture or peasant life. The so-called Bronze Age was really a minor graft onto the still dominant era of stone, wood, and bone.[17] Not really till iron came on the scene, again in serendipitous violence to the canons of materials set by copper (for its cast version is not malleable), was stone really replaced by another material, namely iron. And one should again keep in mind that iron was so flanked by bricks and pots and plasters and cements and glass that its cutting edge in history is often less than we give it credit for. But iron did first dent the world of the farmer and housewife (beyond that of the warrior), through its application to knives and axes and plows.

The path to metals (and what we speak of as the Age of Pyrotechnology) was smoothed by certain basic geological and physical as well as chemical facts. In appearance and abundance, silicon based rocks are the most obvious phenomenon for toolmakers seeking to exploit the surface of the earth (silicon constituting 25 percent of the crust as against 8 percent for aluminum and 5 percent for iron). Had men remained dependent only on muscle power,(viatechnology or the stress shaping of materials), silicon in its rocky form would have remained the dominant material, along with the living materials bone and wood.

The coming of pyrotechnology not only made it possible for men to aggregate common siliceous clays back to a stony shaped material called pottery or bricks (in a manner analagous to the annealing of stone), it also brought into play the basic heat of formation of oxides.

The central ingredient in metallurgy was not the temperature of the melting of metals (except perhaps for copper), but of the reduction of oxides. We have tended to put things backwards in looking at the stability of the metal rather than of its oxide. For the first exploited metals were those that have unstable oxides amenable to reduction by CO at low or moderate temperatures. Silver and gold oxides are decomposed by heat alone.

Iron came on the scene in the form of hematite, limonite, and goethite fluxes for the silica gangue in copper and lead ores. In time iron came to be reduced rather than the copper or lead it was supposed to flux.[18]

Simply by the application of too much iron ore flux and charcoal and blast, iron was produced in bears or salamanders. It was thus discovered to be a useful metal. Alumina-bearing rocks played a fluxing role but never achieved a metallurgic one because of the high energy of formation of aluminum oxide (meaning that to reduce it required unachievable temperatures). Its main early use, therefore, like that of calcium, was confined to cements.

If there is any basic dyad of early pyrotechnologic materials it is iron and silicon, simply because of their back-to -back roles in the accomplishment of iron, slags, and glass. Iron and sulfur (in the form of gossan pyrites) had somewhat less importance.

The dominating concept in early metallurgy thus became one of "polymetallism." Materials were necessarily heated in juxtaposition, as when an iron frit or lead glaze were applied to a clay pot. The

[17] Chêng Tê K'un. 1963. *The Archaeology of China.* (3 vols. Cambridge, England: W. Heffer and Sons), vol. III. *Chou China,* p. 246.

[18] T. A. Wertime. 1968. "A Metallurgical Expedition through the Persian Desert," *Science*, 159:927-935.

juxtaposing of materials was *not* merely the accidental act of men dropping a piece of copper ore into a pottery kiln. It was not even the deliberate effort to extract more energy from a given heat by carrying on two operations at once, such as the kilning of lime and the making of pots (which was almost sure to yield an accidental glaze). It was rather the functioning of a series of technical imperatives that vectored men inevitably towards a common base in the four materials: clays, plaster-cements, iron (as the dominant reducible metal in the earth's crust), and glass.

If any civilizations failed to achieve these four (as the New World failed to achieve cement, glass, and terrestrial iron), they failed because of a negative rather than a positive choice: an option *not* to go further than they did. Critics who argue cultural preferences against the technical sequences necessarily built into fire-shaped materials often ignore this fact. If Chinese civilization chose cast iron, it did so knowing wrought iron but finding the latter uncongenial to its superb casting instincts.

The Old World certainly understood gold-copper *tumbaga*: indeed all the alloys, ores, and metals familiar to the New World. And more. It chose the less exotic metals and materials that lent themselves to a more labor saving and rounded pyrotechnologic existence. To the extent that negative choice functioned (as it did in eliminating *all* rotary motion from the New World but the wheeled toy), Cyril Smith's dictum of curiosity and esthetics as the engine of technical innovation must be modified. Men not only neglected to make the obvious adaptations of well-known materials at crucial thresholds in history. More, some perceptible nooks and crannies they did not want to explore at all, possibly because their particular combinations of raw materials or esthetic interests did not push them in this direction.

The first imperative leading all men toward a common pyrotechnologic experience was that of the oxidizing and reducing and heat altering capabilities of any fire, including of stone annealing fire. Iron turns red (or black, depending on reducing conditions), limestone white, clay and stone darker. Some things cracked or turned powdery, others hardened, still others in time glazed or melted. The identification of these things occupied most of the so-called Stone Age, always against the background set by volcanic glass.

The second imperative was geologic: the stratification that had put stones, clays, and ores into the particular association they possess in the earth. Iron, copper, lead, silica, and limestone are found in common apposition in any mine, especially at contact metamorphic zones. They thus formed naturally interfluxing agencies. Arsenic and tin occur less frequently in these contexts, but sometimes in a pyrite setting. Two of the most basic associations are those appearing in pillow lavas or a sulfide zone of ores; and the subsequent oxidizing forces revolving about the breakdown of chalcopyrite through iron bacteria. Limonite ultimately migrated upwards to the surface of a deposit (along with varying amounts of ores of tin, lead, gold, arsenic, etc.). Copper sulfate, of course, filtered down to an enriched zone, depositing its burden of copper there.

The history of metallurgy revolves in many ways around the ineluctable character of the "gossan," the orange-colored iron caps over the copper (most completely developed in climates that were becoming drier). It was in such settings that Paleolithic men had their first contacts with paints and colors; and Neolithic men their first associations with the capabilities of the gossan to produce interchangeably copper, lead, and iron. At Rio Tinto men could even have developed a natural bronze out of the associated tin.

Those familiar with the gossans of Cyprus would be aware of a ferociously complex chemistry incorporated into so-called Phoenician and Roman slags.[19] They would have observed the capacity of the iron oxide in the gossan to plate surrounding limestones, and of copper sulfate to plate everything in its path including mining timbers.

Sulfide (or sulfate) metallurgy flows ultimately from a gossan—but only through a series of prior steps best characterized not as purely metallurgic but as broadly pyrotechnologic. Little wonder that Cyprus

[19] Arthur Steinberg and Frank Koucky. 1974. "Preliminary Metallurgical Research in the Ancient Cypriot Copper Industry." in L. E. Stager, *et al.*, editors. *American Expedition to Idalion, Cyprus*. Cambridge: American Schools of Oriental Research, p. 158ff.

may have been the terrain for the first true application of iron to the total schema of agricultural life! It was the home *par excellence* of pyrotechnology. Hector Catling has argued that it saw the first true applications of bronze to the domains of the farmer and the housewife.[20]

The third imperative derived from the nature of fluxes, without which reduced metal, even though molten, would remain largely entrapped in the infusible silica gangue of the ore. Fortunately silica forms lower-melting mixtures with most metallic oxides, including those of iron, calcium, and the alkali metals. These relationships between the most abundant oxide minerals in the earth's crust set the stage for much early pyrotechnology. It is also significant that liquids containing large amounts of silica do not crystallize readily on cooling but form glasses that remain viscous but stable over a wide range of temperatures.

A fourth and final imperative concerned the retention of impurities in the melting or smelting operations, impurities that often had a beneficial effect upon the metal. Arsenic and antimony were natural impurities in smelted copper. They educated men towards the seeming superiority of the alloy tin, for bronze. Experience with metal frits and galzes on pots—such as iron and lead and tin—led artisans ultimately to the metal colorants in glass—lead, antimony, iron, copper, and manganese. Even alloys with iron and steel were not far removed. The Chinese lore of gray cast iron came about in some part because phosphorus and silicon were absorbed into the iron, giving it superior casting qualities.

These imperatives reflect the environmental, geologic, thermodynamic, and physical conditions inevitably pushing men by definite phases toward both a pyrotechnology and a metallurgy, both most completely accomplished in the old Middle East. By one token it can be said that men everywhere could not have strayed too far from their influence. This is another way of suggesting that cement, glass, and iron were first discovered in the Middle East and Eastern Mediterranean and could well have migrated by diffusion over the world.

Alternatively, the imperatives were selectively at work elsewhere in the world, driving the first non-Middle Eastern artisans towards the negative cultural choice elaborated in earlier pages of this chapter, including acceptance or denial of stimuli received subtly through diffusion.

Technical imperative and serendipity thus performed together and reinforced each other. They pushed men inexorably beyond the mud brick village towards the world of concrete and steel and glass in which we live. They have brought on the third energy crisis, one that now threatens to terminate the Age of Pyrotechnology and consume us all. This crisis is the imminent exhaustion of the world's fossil fuels. They have helped to introduce a plethora of pollutants to the world, ranging from slags to sulfur dioxide to flourides.

I close with two episodes of personal discovery in the Taurus-Zagros plateau which illustrate the functioning of imperative and serendipity.

In 1966, while at the site of Kuhbanan in the Kerman province of Iran, Cyril Smith and Radomir Pleiner and I came across piles of burned clay bars at a smelting site, theretofore unseen phallic symbols which we labelled "fire bars."[21] Kuhbanan had been made famous by Marco Polo in describing the making of steel or "ondanique" and the capture, on the grates of lead furnaces, of zinc oxide or calamine (named historically cadmia or tutiya). We thought little about Marco Polo when first visiting the mine and seeing how the ancient miners had cut out the galena (lead sulfide ores) while leaving the sphalerite (zinc sulfide) intact.

However, the discovery at nearby Savand, and in 1968 at Ozbekuh, of piles of clay smelting bars began to resurrect Marco— as well as the Arab geographers. Two happenings brought his description home with a vengeance. One was in seeing a small boy apply a bar to a sore on his face. The other was a smelting experience of our own in 1966, in which we tried near Kerman to produce iron in a small lead

[20] Hector Catling. 1964. *Cypriot Bronzework in the Mycenaean World.* Oxford: Clarendon Press, p. 78ff.
[21] Footnote 18.

assay furnace, yielding instead a white calamine fume on the lip of our furnace. To identify the calamine as Marco's *tutiya* involved us in a detective operation.

It became clear in time that the zinc content of the lead ores was not only basically affecting the production of lead (and/or iron from the iron ore flux). What had occurred serendipitously for us was the production of a fume of zinc oxide out of the top of the furnace. The clay fire bars were "grates" for collecting the calamine actually described by Marco Polo. We had rehappened upon a pyrotechnical discovery that was implicit in the character of the ores at Kuhbanan and of the clay grates, but made understandable not only by technical analysis but by the invaluable earlier visit of Marco Polo to the site.

A similar episode on the Black Sea at Trabzon brought another team of mine to an awareness of the famous black sands of the Black Sea coast of Turkey and invoked an earlier presence of Aristotle's pupil, Theophrastus.[22] At Trabzon we came upon an expanse of black sands on the beach, theretofore unreported in most historic accounts of the area. The immediate reaction of the team's geologists, ceramist, and metallurgists was that the black sands, coming from the granites of the Black Sea mountains, could conceivably be a repository for tin for the early Bronze Age.

Further investigation of the sands, carried into the mid-1970's, showed that they grew particularly rich toward Ordu and extended along the whole coast toward Istanbul. But there was no cassiterite, not even a grain, in any analysis. Instead the sands proved to be high in magnetite (up to 75 percent) and olivine, offering a self-fluxing source of iron not unlike that used for the *tatara* furnaces in Japan. They seemed to have provided one attraction for the famous iron-making tribes of Pamphlagonia on the Black Sea, notably the Chalybs. On this subject a wealth of historical record came to play, including Xenophon and Strabo, who viewed the Black Sea as the birthplace of iron, and Homer, who called it the birthplace of silver. Even the 16th century Turks, under Suleiman the Great, had smelted these sands.

But a quotation attributed to Theophrastus offered a broader suggestion as to their role in steel-making.

> "It is said that the production of Chalybian and Amisenian iron is very peculiar; for it grows together, as they assert, from the sand that is carried down by the river. Some say that they simply wash this and smelt it in a furnace; but others, that after frequently washing the deposit left by the first washing they burn it and insert what is called the fire-proof stone which is abundant in this country. This iron is far more beautiful than the other kinds; for if it were not burned in the furnace it would not differ at all, it appears, from silver. Now they say that it alone is not liable to rust; but that it is not very plentiful."

If, in fact, the pattern welding of swords can be traced back to Anatolia of about 1000 B.C., the Black Sea grows in importance as a point of serendipitous innovation in pyrotechnology, even if, in the end, it yields to Cyprus as the place in which iron was brought into play on a major scale as a civilizing element.

[22] To be elaborated in *The Coming of the Age of Iron*.

Flint Technology and the Heating of Stone

by

Jeremiah F. Epstein

Introduction

This paper is concerned with the procedures used to modify flint, chert, and other siliceous stones by the application of heat. It was a process used in many parts of the world by aboriginal flint workers, and became extinct as stone tools were gradually replaced by those of metal. The process was "rediscovered" only recently by Don Crabtree, an expert flint knapper, whose great skill and scientific approach to the subject have been instrumental in reawakening academic interest in the manufacture of chipped stone tools.

Crabtree's discovery of how stone was heated to make it easier to flake was not the first report on the use of fire in flint working. There are a number of accounts of aboriginal peoples in both the Old and New World, who have used fire in quarrying (Holmes 1919), reducing cobbles to flakes (Powers 1877:104), or flaking hot stone fragments by the careful application of water (Lehmann: in Wallace and Hoebel 1952:105). The latter was reported by observers who claimed to have viewed the process themselves. The technique used by the Seri of Sonora, Mexico is perhaps the one most cited:

> "I watched this particular artist for several hours until he had completed an arrow-head that now reposes in my desk, and here is a description of the manner of his labor. Putting three small pieces of flint among the coals of a hot fire on the ground, he places a small stone basin containing a little water within his reach; beside this are placed several straws or reeds of different sizes, together with a few smaller stems of native grass. Presently the first piece of flint placed in the fire is dragged out upon a flat stone by means of a hooked stick, and as the end of the larger straw or reed is dipped in the basin, it will be observed that a drop of water clings thereto; this is lightly touched to the thoroughly heated stone and a small chip flies from the surface. This performance is repeated with astonishing rapidity, until the stone refuses to respond to the touch, when it is returned to the fire and the second stone is treated in the same way, the chips always flying fast and furious. As the work progresses, and the stones are reduced in size and begin to assume the required shape, smaller straws are used, until the final pointing, sharpening, and smoothing is done with the small grasses that pick up a very tiny drop of water and safely remove a very diminutive chip (Holmes 1919:365)."

Holmes' was sceptical.

> "That several hours should have been consumed in the work is difficult to understand, considering the rapidity with which the work was carried forward. A much shorter time would ordinarily be required in producing a like result with hammerstone and bone flaker. A small arrowhead can be made from good quality stone in a few minutes, while larger ones usually require a somewhat longer time (Holmes 1919:365)."

Clearly Holmes' doubts were justified. Accomplished flint knappers have tried the process, without success. Ellis, who directed a laboratory devoted to the study of flint knapping, observed:

> "Experimental attempts to duplicate this fracturing technique have shown it to be very unsatisfactory. In the first place, flint exposed to an open flame for even a short length of time will heat through and shatter into angular fragments and tiny flakes. The portions of the flint which do not shatter may be treated with cold water dropped from the end of a stick or by other means with little reaction. The usual result of the application of cold water to hot flint is the boiling and rapid evaporation of the drops of water. Occasionally small chips may fly off, but their direction and position cannot be controlled. Dropping thoroughly heated flint into a pan of cold water will simply remove the tiny fragments which are broken away by the sudden change in temperature, but often even this treatment has negative results. Also, the flint which has been subjected to fire is so filled with tiny fire cracks and the surfaces of the material so roughened due to the differential expansion of the crystals caused by the heating, that it is impossible to use it to any practical advantage in the shaping of stone implements. An examination of thousands of specimens in the museum failed to indicate that fire played any part in their manufacture (Ellis 1940a:43)."

Earlier, Pond (1930:25) came to the same conclusion, as did Mewhinney (1957:61-2). The fire/water process made no sense, and in many ways served as a red herring in our understanding of how fire could be used in tempering flint. Strangely, until Crabtree's experiments were published, the reports that describe what we know now as correct procedures received very little attention.

If it is proper to talk about landmarks in the rather specialized discipline of lithic technology, one occurred in 1964 when Don E. Crabtree described his experiments and observations on the heat treatment of silica minerals (Crabtree and Butler 1964). Crabtree pointed out from his own experience in flint knapping that while some materials, such as ignimbrite (i.e. welded tuff), obsidian, opalites, jaspers, and agates are easily pressure flaked, the courser microgranular silica materials, such as most chalcedonies, cherts, and flints are difficult to pressure flake in their native state. However, when these are properly heat treated they can be worked with ease. Crabtree's discovery came about while he was working in the Ohio State Museum lithic laboratory in the thirties. Here he observed that the beautiful, multicolored "Flint Ridge" flint used by Hopewell culture (ca. 400 B.C. - 100 A.D.) had a greasy luster, while the quarry material lacked this quality. Later he found the same held for the jasper/agates of Idaho. One fortuitous discovery consisted of an untreated core and a series of flakes that had been taken from it. The flakes, showed the characteristic greasy luster, whereas the core did not.

Crabtree suspected that the lustrous quality of the material may have been due to heating, and so he experimented with the native stone using a sand bath heated by a bunsen burner, or an old coal wood stove. He finally graduated to a ceramic kiln. Through repeated experimentation he discovered that

> ". . . silica materials varied considerably in the length of time and amount of heat necessary to bring about the desired change. For each type of silica material there appeared to be a critical temperature range below which, regardless of the length of time involved, no change

would take place, and above which it would crack and craze. On the other hand, some of the minerals had to be held in the critical temperature longer than others in order to bring about the desired change (Crabtree and Butler 1964:2)."

Crabtree also noted that sudden changes in temperature resulted in the piece cracking or breaking, and it was also observed that thinner pieces such as flakes or roughed out blanks can be treated more successfully than thicker ones.

What is it that makes the flint lustrous after heat treatment? Crabtree examined both untreated and treated specimens under an electron microscope and found that the crystals in the treated pieces were smaller. He interpreted this to mean that heat treatment causes recrystallization of the coarser granular silica materials into smaller crystals, which produced the optical qualities he characterized as a greasy luster. Recrystallization also increased the elasticity of the material.

One problem is how to determine whether strange pieces, from unknown quarries, have been heat treated. The lustrous quality may also be characteristic of some unheated stones, and distinguishing natural from heat induced luster may not be easy. We are told that optical examinations of heat treated pieces made by Prof. Isotoff, Geology Dept., Idaho State University, showed the presence of relic area, i.e., microscopic islands of original crystalline structure. How universal this phenomenon is, is yet to be determined, but . . .

"it does appear to hold promise as a means of identifying heat treatment in unknowns and may well supplement my rule of thumb with respect to luster (Crabtree and Butler 1964:2)."

Since Crabtree's seminal paper, there have been at least two M.A. theses, one Ph.D. dissertation, and a series of shorter papers dealing with heat treatment. In general, the literature has developed along the lines set out by Crabtree. They deal with topics such as: reviews of older sources for better descriptions of heat treatment; the archaeological discovery of pits where material was heat treated; the identification of heat treated stones in various parts of the world; the problems inherent in identification of heat treated stones; and various experimental studies which use scanning electron microscopes as well as standard optical equipment to describe exactly what happens to different kinds of stone when they are heated. The major discoveries are discussed below.

Differences between Heated and Unheated Flint

A number of changes occur in flint when it is heated. The most obvious is that some of the water of hydration within the flint is driven off, and this is detectable by weighing the sample (Purdy and Brooks 1971). The water lies within the interstices of the cryptocrystalline particles, and is probably the cause for the potlidding, or explosions that occur within flint when it is heated very quickly. The same occurs when damp clay is heated rapidly.

Other modifications are best determined by electron microscopic examination. Here a comparison between unheated and heated stones indicated, as Crabtree originally stated, that the heated stone has a less granular structure, or the granules have become much finer. Examination of flints from Florida suggest the impurities found in the intercrystalline spaces of the chert probably act as fluxes when their melting point is reached, changing or transforming the microcrystalline quartz structures to a non-crystalline, amorphous form (Purdy and Brooks 1971:323). Arkansas novaculites seem to have an agranular structure with an increased density of microfractures (Flenniken and Garrison 1975:128). The surface of flake scars in cherts from Kentucky are described as visibly smoother in heated specimens (Collins and Fenwick 1974:138). While there is agreement that water is driven off in the heating process, and the granules of crystalline materials are modified by becoming either smaller or more compact, there are enough variations in flints, cherts, and novaculites so that more specific generalizations cannot be made at this time. It must be emphasized, however, that these changes produce a stone that is chemically no different than it was prior to heat treatment.

In an important series of determinations, Purdy and Brooks measured both the compressive strength and the point tensile trength of treated and untreated Florida cherts. They found that the untreated cherts withstood compressive strength forces ranging from 3000 to 3600 atmospheres, whereas the heated samples could resist forces 25 to 40 percent greater. In contrast, when it came to measurements of point tensile strength, which is most relevant to either percussion or pressure flaking, the heated samples could only withstand forces of 100 atmospheres as opposed to the untreated pieces which were able to withstand 180 atmospheres. This is a precise way of determining that untreated Florida chert is 45 percent[1] harder to flake than heat treated chert. They explain the difference between the two measures as follows:

> "This seeming paradox is easily explained. The binding of the microcrystals which occurs when the rock is heated adds compressive strength through cohesion. The increase in homogeneity which increases strength under compression is the very factor which decreases point tensile strength: (i) the individual microcrystals are bound more firmly together; therefore, (ii) when the flaw is introduced which is preliminary to and necessary for fracture to occur, (iii) failure takes place more readily because the specimen-like glass rather than rock-like aggregate (Purdy and Brooks 1971:324)."

In an interesting demonstration of the difference in flaking quality of treated and untreated Nehawka Chert, Flenniken flaked a series of untreated and heated artifacts, and measured the preforms as well as the debitage afterwards. There was a slightly higher proportion of longer flakes (5.0 mm) from the treated sample (64) than from the untreated or control group (58) (Mandeville and Flenniken 1974:148). Longer, larger, and better controlled flakes were also reported for experiments made on Arkansas novaculite (Flenniken and Garrison 1975:129).

It would seem that as a result of heating, the chert is more homogeneous. This tends to minimize whatever flaws are in the stone, and flakes that might normally break off when hitting flawed areas now carry through (Solberger and Hester '972:182).

Modern and Aboriginal Methods

The experimental work done by Crabtree and those who followed him was done under laboratory conditions. Crabtree's original experiments consisted of putting flakes of various kinds of siliceous materials in a pan of clean sand, placing it in an oven and slowly heating it for days or weeks at a time. He noted that:

> "For each type of silica mineral there appeared to be a critical temperature range below which, regardless of the length of time involved, no change would take place, and above which it would crack or craze. Any sudden raising or lowering of temperature usually ended in extensive crazing or cracking of the mineral specimens (Crabtree and Butler 1964:2)."

Purdy's experiments on Florida cherts were carried out with a kiln, in which the temperatures were elevated by 50 degree increments and held there for 24 hours before being raised to the next level. She discovered that for Florida cherts, the critical temperature was about 350 to 400 degrees centigrade

[1] Sollberger estimated a 30% reduction in effort in working heat treated cherts from Texas (Sollberger and Hester 1972:182)

(Purdy and Brooks 1971:323) when there is an increase in the amount of water lost, and the lustrous appearance develops on the surface. In a later paper, Purdy observed:

> "My first impression . . . was that while early flintknappers were successfully altering lithic materials prior to final retouch, they had only to be concerned with slowly reaching and maintaining the critical temperature (350-400 degrees centigrade). I did not think it mattered if the rock was exposed to temperatures greater than 400 degrees centigrade as long as the exposure was gradual. It is apparent, however, that failure occurs if the rock is subjected to temperatures much higher than necessary for alteration . . . thus, man needed to be very precise in his utilization of fire to alter lithic materials. Not only that, it appears that this precision differed from region to region and depended upon the type of flint materials available (Purdy 1975:140)."

Thus, Purdy's precise laboratory experiments verify Crabtree's earlier conclusions.

How was heat treatment accomplished by the aborigines? Here we are fortunate in having some ethnohistoric descriptions as well as archeological discoveries as guides.

For the Plateau Shoshone, there is a description given us by Powell (1875:27-28).

> "The obsidian or other stone of which the implement is to be made is first selected by breaking up larger masses of the rock and choosing those which exhibit the fracture desired and which are free of flaws: then these pieces are baked or steamed, perhaps I might say annealed by placing them in damp earth covered with a brisk fire for twenty-four hours. . . .

Another, cited by Hester (1972:64), describes a practice of one of the northern Plains Indians:

> "The material used by one of the men is a black obsidian obtained by trade from the Crows to the south, while the other has a piece of milky chalcedony picked up in the mountains to the west. Each of these blocks has been sweated by being buried in wet earth, over which a fire has been built, the object of this treatment being to bring to light all the cracks and checks in the stone so that no unneccessary labour need be performed on a piece too badly cracked to be profitably worked (Grinnell 1898:142)."

In both accounts a fire was positioned over wet earth surrounding the flint. Presumably damp soil keeps the flint from being heated too rapidly. But the reports differ in the function of the fire. Powell believes that heat treatment was for annealing flawless stone, whereas Grinnell specifically states that the purpose of heating was to detect flaws. It is hard to believe Grinnell is correct, since the easiest way to check for flaws is to work the stone first. I, therefore, believe he was either deliberately misinformed, or misinterpreted what he was told.

Neither of these accounts involve complicated procedures; mounding the damp earth and a campfire are sufficient. Experiments made in California indicate that an ordinary wood burning campfire can raise the temperature of stones lying on the ground to 1000° F. (Harner 1956:40).

At least one archaeological discovery indicates that minor excavations were made for the express purpose of heat treatment. At the Spillway Site in Kansas, a pit about 2 1/4 feet in diameter, and about 1 3/4 feet deep was dug into the soil. The bottom was covered with a 4" thick bed of ashes, which was in turn covered by a 4" thick layer of flint flakes and crude blade cores spread evenly over the ashes.

These were capped with three limestone boulders and, presumably, soil. In spite of the fact that the ashes suggest the fire of considerable intensity, the flint showed no sign of having been subjected to a blazing fire (Shippee 1963:271). Unfortunately, Shippee's report fails to note whether the flint had the lustrous quality that characterizes properly tempered flint. Shippee also noted that the well known flint knapper Marvin McCormick

> ". . . found it necessary to 'temper' much of the material he used by burying it and building a hot fire above the deposit, then letting it cool slowly. Some flints needed more than one heating (Shippee 1963:271)."

Purdy (1971:13-14) collected a number of late 19th century reports that describe caches of unfinished flint artifacts in contexts which strongly suggest that they had been intentionally heat treated. Two of these are of special interest:

> "Dutchess County, New York: While employed in digging his spade brought up a number of arrow-points. He described them to be nicely piled side by side, edgewise, in two or three rows. There were perhaps two or three hundred in all. On each side and on top were some charred logs and sticks, that seemed to be the remains of an old fire. They were 10 or 15 inches below the surface of the pond. They are of a blue jaspery flint, and seem to be in an unfinished condition. I thought that probably the Indians had brought them from a distance (as I have never found any of the same rock anywhere in this neighborhood) and made this pocket and covered the traces of them by building a fire, intending to return and finish them at their leisure; or, perhaps, they hid them there to prevent their capture by their enemies (Shepard 1877:307-8)."

> "In 1894, Dr. J. R. Snyder excavated Mound No. 1 of the Baehr Group, on the west side of the Illinois River, thirteen miles below Beardstown, and opposite the mouth of Indian Creek (Illinois): At the base of the mound was an oval of clay on which "was a mass of black hornstone implements, that apparently had been thrown down in lots of 6 to 20, with sand over and between each lot, as though to isolate them from each other. This deposit of 6,199 flints was covered with a stratum of clay, 10 inches in thickness; and on this a fire had been maintained for some time. . . . The flints forming the nucleus of this mound are very . . . rudely fashioned; some are quite neatly finished, but the greater part of them are only chipped and ill-shaped . . . they do seem to be unused blanks (Ellis 1940b:112."

An interesting and successful attempt to temper flint using a pit, during the winter when the ground was cold, was carried out by Mandeville and Flenniken (1974). Theirs, which was roughly 70 cm. in diameter and 60 cm. deep, had thermocouples placed at various locations to check the temperature. The pit was first heated for two hours until the coals reached a depth of about 35 cm. The coals were covered with another 6 cm. layer of sand and flint preforms over which was built another fire which also burned for two hours until it had a 35 cm. thick layer of coals. The pit was then covered with 10 cm. of sand.

> "The temperature among the preforms rose from 75 to 300° C. in an hour and a half and then began to descend rapidly. At this time the fire was capped with sand; the temperature rose precipitously to 325° C., where it remained for an hour and a half before beginning a slow and gradual descent. Twenty hours later, when the temperature had fallen to 75° C., the preforms were removed (Mandeville and Flenniken 147)."

As noted earlier, the flint so heat treated flaked better than untreated samples.

In summary, the data are far from consistent. Fires are made either under or above the flints, or both, and the stones are sometimes in direct contact with the ash (i.e., fire), and at other times insulated by damp soil. Some pieces have to be heated more than once. The laboratory studies (using controlled heat in kilns) of Crabtree, Purdy, and Flenniken and Garrison all emphasize slow gradual heating. Yet in Mandeville and Flenniken's pit, the temperature rose from 75 to 300 degrees centigrade in an hour and a half, apparently without injuring the flint.[2] Both Purdy and Crabtree emphasize that the tempering temperatures have lower and upper limits to be successful, and yet the examples from the historical literature, as well as those seen archaeologically, suggest that narrow temperature controls were not achieved. There appears to have been considerable variation in the way that flint was treated and, along with this, a significant variation in success rates. It is hard to believe that the aborigines expected perfection in their tempering processes. I suspect they stacked their pits with the expectation that some of the artifacts would temper well enough to serve their purpose. Those that were not fired sufficiently could always be reheated.

Intentional Heat Treatment in an Archeaological Context

At this time the best clue to heat treatment is what Crabtree defined as a greasy luster, which is visible on the surface of the piece. In an old report on the Andaman Islands where heat treatment is mentioned, we are told in a footnote that they used a bluish white quartz with a *fatty lustre* (Man 1883:380). In almost all of the recent studies luster is mentioned, though often without an adjective. Occasionally the word "vitreous" is used. While it will vary from stone to stone, it is usually obvious enough to be visible. We may also find some alteration of color if the flint has sufficient iron compounds in it, in which case the tempered stone will change to various shades of red. Stones that are overheated will show the characteristic oval "pit-lids" and crazing that have been recognized as features of burned flint, as opposed to properly tempered material. Accurate diagnosis probably requires having the unaltered source material available for comparisons, since some flints appear to have a lustrous quality in their natural state. Electron scanning microscopes or high powered optical equipment available, can be exceedingly helpful.

Once a treated piece is identified, how can we determine whether this alteration was purposeful? There are so many situations in which stone can be heated without the intention of tempering that distinguishing one from the other seems an impossible task. Fires are often built over a bed of cobbles, heated stones are used in cooking, either in earth ovens or in stone boiling. If these stones were siliceous they would be altered. Then too, wherever man has built fires we find he often builds them over his own rubble, which usually consists of the remains of his flint working as well as his other activities. There is ample opportunity in the cooking/camping milieux to heat siliceous stone or stone fragments without consciously trying to temper them.

Quarrying also involved fire. For example, in Ohio, the Hopewell people who used the beautiful varigated stone from Flint Ridge found it from 10 to 20 feet below the surface. They had to break through a bed of limestone before reaching the flint deposit. It appears that fire was used to heat the limestone, and water was then covered with clay to protect it (Holmes 1919:176). Fire was apparently also used in quarrying for the novaculites of Arkansas (Holmes 1919:198). I would assume that the fire used in quarrying would also temper, to some degree, the flint that was quarried. In this connection, it should be rembered that Crabtree's first awareness of heat treatment was in connection with the Ohio Flint Ridge material, and it has been demonstrated that the artifacts of Arkansas novaculites were also tempered (Flenniken and Garrison 1975). In these instances it perhaps doesn't matter whether the stone was heated to improve its flaking qualities or just to mine it. What is important is that the tempered stone was used, and presumably its qualities were appreciated.

As of this writing, the only study specifically concerned with the problem of identifying archaeological sites where stone was intentionally treated, is that of Collins and Fenwick (1974). They examined

[2] Sollberger achieved an even more precipitous rise in temperature without injuring the stone when he successfully tempered the tips of projectile points by placing them in a butane flame for 30 seconds (Patterson 1975:10)

chipped stone collections from twelve different sites in Kentucky. Since they do not mention earth pits or ovens I suspect these features were not found. Four criteria are employed to identify heated stones: (1) color change; (2) heat damage; (3) luster, and (4) a sugary texture which appears on the interior of flakes from Western Kentucky cherts, that had been identified in heat treating experiments (Collins and Fenwick 1976:147). They are, therefore, dealing with all indications of heating, rather than just luster. The report, while interesting, is marred by inadequate data and conclusions which depend upon information not fully supplied. Their only table lists the sample size and the number and proportion of heated implements and debitage from twelve different sites in Kentucky. Not given are the numbers and percentages of the various classes of artifacts that show (or fail to show) heat treatment (points, scrapers, etc.) or the numbers/percentages of artifacts and debitage that show each of the four features used in diagnosing heat treatment. These drawbacks limit the number of specific questions about heat treatment that can be asked . For example, the authors conclude that one indication of intentional treatment is where 20% or more of the implements exhibit thermal alteration in one form or another, for this percentage is considered higher than would be expected by chance. Why they choose 20% is not at all clear. I have rearranged their data to show the number of sites and the percentage of heated implements in them. The break down is as follows:

Number of Sites	% Implements Heated
1	less than 10
4	10 - 14.9
2	15 - 19.9
3	20 - 24.9
0	25 - 29.9
0	30 - 34.9
2	35 - 36.8
12	

As I see it, there is no basis in distinguishing intention from accident until one group is separated from the other which occurs at 35%—not 20% as stated by Collins and Fenwick. This suggests that intentional heat treatment occurs at only two sites.

The authors also observe that there was a preference at most sites for heating flakes or bifacial preforms and consider this to be an indication of intentional heating. This is obviously a crucial point, and is what we should expect if heating was used intentionally. But we would like to know how many pieces out of all of the flakes and all of the preforms show heat treatment, and this information is not supplied. Also lacking is data on the proportion of pieces showing luster. In one instance only, Bland Cave, is it possible to figure out that about 7% of the total sample of 2515 pieces have this quality. In other examples, we are informed that only a few pieces show luster or other qualities of heating. From this one may assume that the feature of luster, which Crabtree originally pointed out as a quality of heat treated stone, is not too frequent in archeological samples--at least in Kentucky.

One of the most puzzling aspects of Collins' and Fenwick's data occurs in regard to the relationship of heated chert in both the debitage and the completed artifacts. Here we see the proportion of heated chert is about the same.

Site	Implements			Debitage		
	Total Sample	Heated #	Heated %	Total Sample	Heated #	Heated %
HO 41	751	277	36.8	1764	689	39.1
LV 27	712	152	21.3	1523	639	39.1
PI 11	235	51	21.7	1247	292	23.4
HK 49	711	80	11.2	8	1	12.5

These figures are intriguing if only because similarities in proportions beg for explanation. Two opposing interpretations seem possible. On one hand I would suppose that if the chert was all accidentally heated, we would expect about the same proportion of heat altered implements as debitage. Yet, if the heating was intentional, and a given percentage of their artifacts were made from heated chert, would we not expect the debitage to reflect approximately the same proportion? This paradox would be resolved if the information on the classes of implements and debitage had been supplied.

In summary, the only study to date that addresses itself to the very important question of intentional versus unintentional heat treatment while not as well thought out as it should be, indicates that in Kentucky, heat treatment was used in Archaic, Early Woodland, and Adena cultures, and possibly in Paleo-Indian times as well.

Distribution of Heat Treatment

As of this moment, heat treated flint has been reported the Solutrean of France (Bordes 1969; Collins 1973), on a fluted point from California (Davis and Shutler 1969), from various complexes in Texas (Hester and Hill 1973; Hester and Collins 1974; Solberger and Hester 1972) Arkansas (Flenniken and Garrison 1975), Kansas (Shippee 1963), Florida (Purdy and Brooks 1971), Kentucky (Collins and Fenwick 1974), Ohio (Crabtree and Butler 1964) and Idaho (Crabtree and Butler 1964). A glance at the names of those who have made the discoveries indicates a rather small group of scholars, who, either directly or indirectly, have had contact with Don Crabtree. It would seem that indications of heat treatment are found wherever there are people qualified to recognize it. The scattered references to heat treatment in the New World suggest that in the United States, heat treatment was used from Paleo-Indian up to historic times. It is probably realistic to assume that in the next twenty years we will document its practice in much of Europe, Asia, and Central and South America. Obviously, if that information is not forthcoming, then we will have to assume that the process was invented independently in both the Old and New World.

The ethnohistoric accounts tell us the process was carried out in various ways from covering the material over with wet clay and building a fire on top of it, to simply thrusting the piece to be tempered in a fire. Even though I suspect, as does everyone else who has been concerned with this subject, that many of these reports are not accurate, they at least point out that heat treatment was not a uniform procedure throughout the world. Therefore, it is not enough just to identify tempering, we must also try to determine how it was accomplished. This means more digging, with more attention paid to hearths, ash deposits, and the distribution of the various kinds of burned stone that occur in most archaeological sites. It also behooves us to return to the older literature for detailed information on the many caches of "blanks" that have been reported. Eventually we may isolate various traditions of heat treatment, each of which utilizes different procedures.

Heat Treatment of Stone in the History of Pyrotechnology

Since our knowledge of the ancient art of heat treatment of siliceous stone is much too meager to permit any valid generalizations about its history, I would like to conclude this paper with some speculations that may be helpful in viewing this process in perspective.

We do not know whether heat treatment of flints and cherts was invented just once, or whether it developed independently many times in many places. Now that archaeologists are alerted to the process it should not be too long before we obtain enough spacial-temporal information to determine if indeed there is or is not a pattern to the history of heat treating stones. At this time the earliest evidence of the process is from the Solutrean in France. In the New World the process first appears in the Paleo-Indian cultures of the terminal Pleistocene. Both the Solutrean and the Paleo-Indian are noted for their mastery of flint working, particularly in regard to the production of large, pressure flaked, bifacial points. Insofar as flint knapping is concerned, both cultures are the products of true craftsmen, and so we can assume that, like craftsmen everywhere, they studied their medium in a deliberate effort to learn its potentialities and limitations. It is in this context that the experimentation necessary for achieving the proper temperature control for annealing flint becomes most comprehensible. The observation that some heated stones flaked better than others was probably made many times during the half million or more years that man used fire. But it appears that these discoveries, did not take root until relatively late in the Upper Paleolithic. The reason seems obvious . . . . annealing flint is most useful when one is concerned with pressure flaking, with the superb mastery of flint that is required to produce an object whose beauty transcends its functional requirements. It is in the Solutrean of France that pressure flaking first achieves this perfection, a perfection that was later paralleled in the Paleo-Indian cultures of North America.

Although we cannot peer into the minds of Upper Paleolithic craftsmen, we can realistically assume that by tempering flint they gained an increasing appreciation of how fire could modify the world in which they lived. By Solutrean times, man probably knew how roots and seeds could be converted into more digestible forms by roasting and how the same process made meats and fish more palatable. The technique of fire-hardening wooden spear tips was already known by third interglacial times in Germany (Mouvius 1950), and it is not unreasonable to assume that man had also discovered that logs could be hollowed by the carefully controlled application of fire combined with assiduous scraping with flint tools. Thus, by the latter part of the Upper Paleolithic, we can be reasonably confident that man had accumulated a fairly large body of lore relating to the ways in which fire could alter organic substances.

What makes the heat treatment of siliceous stones so significant is that it represents, to our knowledge, man's earliest deliberate effort to modify, by heating, *inorganic* substances. Much later this was followed by the firing of clay to produce pottery, and finally by metallurgy.

It is always tempting to view a historical sequence as an evolutionary development with each step the necessary precursor to successive ones. If we think in this way, we cannot help but wonder whether the heat treatment of flints had implications for later technological innovations in man's rise to Parnassus. Did the skills involved in controlling fire for the tempering of siliceous stones ultimately take root somewhere and lead to the development of pottery? First, of course, we need to know whether flints or cherts were annealed in those parts of the world where pottery first appears . . . such as in Iran, or southeast Asia. What is perhaps most exciting is that a speculative question of this nature is answerable once the data is in. Now that the question has been posed, I suspect that the answer will be soon in coming.

References

BORDES, F.
1969 "Traitement thermique du silex au Solutreen." *Compte Rendus des séances mensuelles de la Société Préhistorique Francaise,* 66(7):197.

COLLINS, M. B.
1973 "Additional Observations on the Thermal Treatment of Chert in the Solutrean of Laugerie Haute." *Proceedings of the Prehistoric Society for 1973.*

COLLINS, MICHAEL B. AND JAMES M. FENWICK
1974 "Heat Treatment of Chert: Methods of Interpretation and their Application." *Plains Anthropologist*, 19(64):134-145.

DAVIS, E. L. AND R. SHUTLER, JR.
1959 "Recent Discoveries of Fluted Points in California and Nevada." *Anthropological Papers, Nevada State Museum*, 14:154-169.

CRABTREE, D. E. AND B. R. BUTLER
1964 "Notes on Experiments in Flint Knapping, 1. Heat Treatment of Silica Minerals." *TEBIWA*, 7(1):1-6.

ELLIS, H. HOLMES
1940 "Flint Working Techniques of the American Indians: An Experimental Study." *The Ohio Historical Society Publication*, No. 72. Columbus.

FLENNIKEN, J. JEFFREY AND ERVAN G. GARRISON
1975 "Thermally Altered Novaculite and Stone Tool Manufacturing Techniques." *Journal of Field Archaeology*, 2(1/2):125-131.

GRINNELL, G. G.
1898 *The Story of the Indian*. D. Appleton and Co., New York.

HARNER, M. J.
1956 "Thermofacts vs. Artifacts: An Experimental Study of the Malpais Industry." *University of California Archaeological Survey Reports,* 33:39-43.

HESTER, THOMAS ROY
1972 "Ethnographic Evidence for the Thermal Alteration of Siliceous Stone." *TEBIWA*. 15(2):63-65. Pocatello.

HESTER, THOMAS ROY AND MICHAEL B. COLLINS
1974 "Evidence for Heat Treating of Southern Texas Projectile Points." *Bulletin of the Texas Archaeological Society*, 45:219-224.

HESTER, THOMAS ROY AND T. C. HILL, JR.
1973 "Prehistoric Occupation on the Holdsworth and Steward Sites on the Rio Grande Plain of Texas." *Bulletin of the Texas Archaeological Society*, 43:33-75.

HOLMES, W. H.
1919 "Handbook of Aboriginal American Antiquities. Part 1. Introductory—the Lithic Industries." *Bureau of American Ethnology Bulletin*, No. 60. Washington.

MAN, E. H.
1883 "On the Aboriginal Inhabitants of the Andaman Islands, Part III." *Journal of the Anthropological Institute of Great Britain and Ireland*, 12:327-434.

MANDEVILLE, M. D. AND J. JEFFREY FLENNIKEN
1974 "A Comparison of the Flaking Qualities of Nehawka Chert before and after Thermal Pretreatment." *Plains Anthropologist*, 19:146-148.

MEWHINNEY, H.
1957 *A Manual for Neanderthals*. Austin.

MOVIUS, HALLAM L. JR.
1950 "A Wooden Spear of Third Interglacial Age from Lower Saxony." *Southwestern Journal of Anthropology*, 6:139-142.

PATTERSON, L. W.
1975 "Selective Heat Treating of Flint." *Texas Archaeology: The Newsletter of the Texas Archaeological Society*, 19(3):10-11.

POND, ALONZO W.

1930 "Primitive Methods of Working Stone: Based on Experiments of Halvor L. Skavlem." *Logan Museum Bulletin*, No. 3. Beloit.

POWELL, J. W.

1875 *Exploration of the Colorado River of the West and its Tributaries Explored in 1869-1872.* (Under the Direction of the Secretary of the Smithsonian Institution). Washington.

POWERS, STEPHEN

1877 "Tribes of California." *Contributions to North American Ethnology*, 3(9):104. Washington.

PURDY, BARBARA ANN

1971 *Investigations Concerning the Thermal Alteration of Silica Minerals: An Archaeological Approach.* Ph.D. Dissertation, Department of Anthropology, The University of Florida.

PURDY, BARBARA ANN

1975 "Fractures for the Archaeologist." In *Lithic Technology: Making and Using Stone Tools*, pp. 133-141. Edited by Earl Swanson. The Hague.

PURDY, B. A. AND H. K. BROOKS

1971 "Thermal Alteration of Silica Minerals: An Archaeological Approach." *Science*, 173:322-325.

SCHUMACHER, P.

1877 "Methods of Making Stone Weapons." *U.S. Geographical and Geological Survey Bulletin*, 3:547-549.

SHEPARD, EDWIN M.

1877 "Deposit of Arrow-heads near Fishkill, N.Y." *Annual Report of the Smithsonian Institution*, 1876, pp. 307-08. Washington.

SHIPPE, J. M.

1963 "Was Flint Annealed Before Flaking?" *Plains Anthropologist*, 8(22):271-272.

SOLLBERBER, J. B. AND T. R. HESTER

1972 "Some Additional Data on the Thermal Alteration of Siliceous Stone." *Bulletin Oklahoma Anthropological Society*, 21:181-185.

WALLACE, ERNEST AND E. ADAMSON HOEBEL

1952 *The Comanches, Lords of the South Plains*. University of Oklahoma Press. Norman.

Addendum

Since this symposium was given, Don Crabtree and Tom Hester have called my attention to other examples of heat tempered flint. This new information, in part, extends our knowledge of the geographical distribution of the process. Heat treatment is now well documented for the Great Basin, and for Arizona, particularly during the paleo-Indian and early Archaic periods. Crabtree reports that he saw heat treated points from Fells Cave, Chile, and from Tikal, Guatemala, and other mataerials came from Sendai, Hokkaido. Apparently modern Australian aborigines also "cook" flint. Thus heat treatment of flint is widely dispersed both in the Old and New World, and it is quite probable that we will soon document its occurrence in the intervening areas. The problem is now to determine how many flint tempering traditions there are, and whether they arose independently.

I am especially grateful to Denise Schmandt-Besserat for pointing out to me that thermal alteration of ochre occurs in the Chatelperonnian of France and in the earlier levels at the Molodova site in Russia. If this small sample of heat treated materials is indeed representative, then the whole process may have started with the production of suitable pigments which then was extended to the modification of flints.

II. COMMUNICATION OF KNOWLEDGE AND TECHNOLOGY

On the Origins of Writing

by

Denise Schmandt-Besserat*

The Present State of Knowledge

The origins of writing are still surrounded with mystery. All we know is that writing first appears in Mesopotamia but the date remains only approximated to 3250-3100 B.C. and it is disputed whether the Sumerians or rather the Subarians (their predecessors in upper Mesopotamia) are the inventors. Writing is assumed to have evolved from a first pictographic stage where each word is represented by a small picture or logogram, to the elaborate cuneiform script where the characters, or phonograms, represent sounds.[1] We base our present knowledge of the beginnings of writing on a body of around 4000 texts recovered not only in Mesopotamian sites such as Warka, Kish, and Tello, but also in the Diyala region: Khafaje; in Iran: Godin Tepe, Chogha Mish and in Syria: Habuba Kabira and Jebel Aruda. The content of the texts is for the greatest part still enigmatic to epigraphists but there is strong indication that they were economic records. The seal impressions they bear confirm their official use.

Some Characteristics of the Early Tablets

The early texts have a number of enigmatic features.

1. The signs belong to two types:

(a) Pictographs which depict in a readily recognizable fashion the objects they stand for. This type is rare and represents vocabulary of infrequent use such as chariot, variety of birds, wolf, etc.

(b) Totally abstract signs which cannot be understood by the non-initiated. This second category constitutes the great majority of the signs (which explains why most of the tablets are still undeciphered). It is important to note that while some characters assume totally odd shapes, series consist of multiple variations of basic forms including circles, cones and rods. (plate VII, ill. 1). Indeed, among 950 signs compiled by Falkenstein from the Warka tablets, 107 (11 percent) are variations on circles, semi-circles, ovals, etc., 96 (10 percent) are variations of cones and diamonds, and 42 (4 percent) are rods of various sizes and bearing different striations and appendices.[2] The abstract signs which have so far been deciphered represent commodities of current usage. For instance the sheep is represented by a circle enclosing a cross, bread as a cone with an incised line parallel to the base, oil as an ovoid with a deep groove around the larger extremity and beer as a cone with a constriction above the base. Numbers are also represented by variations of cones and circles which are differentiated from

* I am indebted to Ellen Simmons for the drawing of all the illustrations, maps and charts.

[1] I.J Gelb. *A Study of Writing*. The University of Chicago Press, 4th ed., 1974. p. 62ff—and David Diringer, *Writing*, Ancient People and Places. Thames and Hudson, London, 1962. p. 36ff.

[2] A. Falkenstein. *Archaische Texte aus Uruk*. Ausgrabungen der Deutschen Forschungsgemeinschaft in Uruk-Warka, Deutsche Forschungsgemeinschaft, Berlin 1936.

the remaining signs by the technique of impressing them with the blunt part of the stylus while the other characters are incised with the pointed tip.

2. The repertory of the first scribes consists of about 2000 signs.[3]

3. Although the sites where the early texts have been recovered are scattered over a wide area extending from Warka in Mesopotamia to Habuba Kabira in Syria and Godin Tepe in Iran, the signs used are mostly identical and evidence a rapid standardization.

4. Except for land sales which are usually written on stone tablets (plate VII, ill. 2), the early texts are written on tablets made of clay—a rather impractical material which smears easily and requires an extended drying or firing process before circulation (plate VIII, ill. 3).

5. The faces of the stone or clay tablets are not flat as one might expect, but are cushion shaped with convex surfaces.

The abstract character of the signs, the extensive repertory of vocabulary, and its immediate standardization seem to indicate an already well developed and sophisticated writing system, not a humble beginning. Some epigraphists have suggested that prior to the tablets an earlier phase of writing may have existed, possibly practiced on a perishable material which has not survived. I am inclined to think that the shape of the abstract signs, the choice of material and the shape of the tablets may be leads to an earlier prototype.

A New Hypothesis

In 1964, Pierre Amiet, conservateur en chef du département des Antiquités Orientales au Musée du Louvre, identified in the archives of Susa an archaic system of recording dating from the second half of the IVth Mill, B.C., and slightly predating the earliest tablets.[4] The system consists of small clay tokens of geometric shapes mostly in the form of spheres, discs, cones, and tetrahedrons, found enclosed in clay envelopes in the shape of hollow clay balls (called *bullae*) (plate IX, ill. 4). The surface of the *bullae* usually bear seal impression and sometimes marks indicating the number of tokens enclosed. Pierre Amiet interprets each *bulla* as representing a transaction, the tokens inside indicating the kinds of goods exchanged by their shapes and the quantity by their size and number.

In the course of my recent study on the earliest uses of clay in the Middle East, I found that geometric tokens identical to those identified by Pierre Amiet in the *bullae* are found loose in most Middle Eastern sites of the IXth to the IId Mill, B.C.[5] I have been intrigued by the striking similarity of the shapes of the tokens with some of the abstract signs from the Uruk tablets. In the present paper I will discuss four stages in the evolution of an early recording system based on tokens which may have led to writing.

1. Ca. 8500 B.C. A three-dimensional system of recording is based on tokens.
2. Ca. 3250 B.C. Clay containers or *bullae* hold the tokens of a particular transaction.
3. Ca. 3250 B.C. Signs are impressed on the surface of the *bullae*.
4. Ca. 3250-3100 B.C. Clay tablets bear impressed and incised signs.

1. The Tokens

Clay begins to be currently used in the Middle East around 8500 B.C. and among the first clay objects are series of geometric tokens including spheres, discs, cones, tetrahedrons, long pellets, as well as odd shapes (plate IX, ill. 5). Those tokens are of small dimension ranging from 1 to 2.5 cm and rarely

[3] *Ibid.*, p. 29.
[4] Pierre Amiet. "Glyptique Susienne." *Mémoires de la Délégation archéologique en Iran*, Vol. XLIII. Paris 1972. Vol. I, p. 69ff.
[5] Denise Schmandt-Besserat. "The Use of Clay before Pottery in the Zagros," *Expedition*, Vol. 16, No. 2, 1974; "The Earliest Uses of Clay in Syria," *Expedition*, Vol. 19, No. 3, 1977; and "The Earliest Uses of Clay in Turkey," *Anatolian Studies*, Vol. 27, 1977.

reaching 3 to 5 cm. Rare examples are reported in stone, ochre, plaster and bone, but the great majority are made of fine untempered clay. Their careful to sloppy manufacture was achieved by simple hand modeling, either by rolling them between the palms of the hands or by pinching them with the finger tips. Their color, ranging from red to buff with many blackish examples indicates that light firing was involved.

Tepe Asiab, Ganj-i-Dareh Tepe and Beldibi are the earliest sites where the geometric objects occur and the three sites may be dated to the middle of the IXth Millenium B.C. (Ganj-i-Dareh Tepe, level E, 8450 ±150 GAK 807). Tokens are present in virtually all Neolithic sites—for instance at Jarmo, Suberde, Anau, Tell Ramad, Jericho and Khartoum; in Chalcolithic sites such as Arpachiyah, Tell as Sawwan and Tal-i-Iblis. They continue in the Bronze Age at Tepe Hissar, Susa, Jemdet Nasr, Warka, Tepe Gawra, Kish, Ur and Chanhu Daro. They are still present in the IId Millennium, for instance at Tepe Hissar, Megiddo (plate X, ill. 6). The list of the sites above mentioned illustrates that they are not only found in a long time period but also in a geographical area of considerable extension, including Egypt, Palestine, Syria, Anatolia, Mesopotamia, Persia, and the Indus Valley (plate XI, ill. 7). The geometric tokens are often discussed in site reports, but they seem to have escaped special attention of archeologists because they are usually reported under various headings. The cones are sometimes incorporated with the figurines, the spheres with "games" and the rest under headings such as "small objects of uncertain purposes." They are never seen as belonging together. The number of the geometric tokens at each site may reach sizeable proportions; for instance, the Jarmo report mentions 1153 spheres, 206 discs, and 106 cones.[6] The geometric objects are usually found in clusters on the house floors and they may have a tendency to be most numerous in storage areas. If they were held in containers, these must have been of perishable material such as textile or skin bags. Their shapes, which are presently of greatest interest to us, are from the beginning numerous. They include spheres, discs, cones, and cylindrical pellets in variable sizes and proportions. Each type may be divided in series of subtypes, such as 1/2, 3/4 spheres, tetrahedrons, pyramids, biconoids, etc. Again specimens of each type and subtype exhibit incised and punched marks which seem intentional and probably bear a meaning.[7]

The definite shape of the tokens, the care with which they are made, their firing, their number and their widespread distribution in space and time suggest that they had a precise function of some importance. What could it be? Oppenheim describes an interesting accounting system used to keep count of the animals in the herds of the palace of Nuzi. Each animal was represented by a token and deposited in a basket. Tokens would be transferred to appropriate baskets to keep track of change of pasture or shepherds, when animals were shorn, etc.[8] This system of accounting seems to have been widespread in Mesopotamia, and Iraqi shepherds today still keep count of their animals with pebbles.[9] It was probably used also to count all possible commodities, and the Bible pictures Yahweh keeping account of all humans with the same device: each living individual was represented by a token in a receptacle. Death was believed to occur when an angel would whirl a token away with a sling.[10] This token system is similar to many simple counting systems: the Romans used small pebbles, "calculi", as counters, and many of us have learned to add and subtract with an abacus. The Nuzi system of accounting described above dated from the IId Mill B.C. but there is no reason not to believe that such a system could have existed much earlier. The system of tokens of the Neolithic period was, however, more elaborate than simple pebble counters as it involved a repertory of numerous shapes. What could have been the meaning of those shapes? It is striking, as stated above, that the tokens have some of the same basic shapes as some of the abstract signs of the early tablets (plate XII, ill. 8). The number one is represented by a conical mark impressed with the blunt end of a stick pressed diagonally against the

[6] Vivian L. Broman. *Jarmo Figurines* Unpublished Master's Thesis, Harvard University-Peabody Museum Library, Cambridge, Mass., 1958, pp. 62, 63, 58.

[7] Denise Schmandt-Besserat. "An Archaic Recording System and The Origin of Writing", *Syro-Mesopotamian Studies*, Vol. 1, No. 2, 1977, pp. 1-32; and "The Earliest Precursor of Writing", *Scientific American*, Vol. 238, No. 6, 1978, pp. 50-59.

[8] A. Leo Oppenheim. "An Operational Device in Mesopotamian Bureaucracy" *Journal of Near Eastern Studies*, Vol. XVII, 1958, pp. 121-28.

[9] Thorkild Jacobsen. "The Relative Roles of Technology and Literacy in the Development of Old World Civilizations", paper No. 13 in Sol Tax, *Human Origins*, an introductory general course in Anthropology, selected readings, Series II, Second Edition, The University of Chicago Press, 1946, pp. 241-49.

[10] Otto Eissfeldt. *Der Beutel der Lebendigen* Berichte über die Verhandlungen des Sachsischen Akademie der Wissenschaften zu Leipzig, Philologisch-Historische Klasse, Band 105, Heft 6, Akademie Verlag, Berlin, 1960.

tablet; 10 is a circular impression impressed with the same tool pressed perpendicularly against the tablet; 60 is a large cone; 600 is a cone with a circular impression; and 3600 is a large circle. We can assume that the tokens in the shape of a small cone were standing for 1 and the small spheres for 10 etc. This would explain the most common shapes of the tokens, but what would the more complicated shapes stand for? Here again, some of the shapes of the signs standing for the most usual commodities can be equated with token shapes. The triangles with 5 incised lines probably stood for "metal, silver"; the discs with an incised cross for "sheep"; the biconoids with a concentric incision for "sweet" and the rectangles with an incised cross stood for "building, room" (plate XIII, ill. 9).

2. The Bullae

Bullae are hollow clay balls which are found totally closed, usually bearing seal impressions. When shaken they produce a rattling noise, indicating some loose content, and when broken, they are found to hold a various number of tokens. The *bullae* appear in a period covering the second half of the IVth Mill and the beginnings of the IIIrd Mill B.C., not only at Susa where they were first identified but also in various sites such as Chogha Mish, Tepe Yahya, Tal-i-Malyan and Shah-dad in Iran; Warka in Mesopotamia, and Habuba Kabira in Syria.

The *bullae* may have been used as bills of lading as suggested by Pierre Amiet, accompanying finished products such as textiles, manufactured in the country and transferred to the temple.[11] The producer consigned his goods to the care of the middle man together with a *bulla* containing the number of tokens equivalent to the shipment. The *bulla* was duly sealed for authentication. Amiet notes that there are usually two different seal impressions,[12] probably those of the sender and the middle man. The recipient of the delivery could, by breaking the *bulla,* check the accuracy of the shipment upon arrival.

The *bullae* may also have been devices to transfer tokens from one account in a particular service of the temple to another, such as described by Oppenheim in Nuzi. In this case, the first seal would represent an accountant and the second, a supervisor or the recipient, such as a shepherd. Accounting device, or bill of lading? The interpretations are certainly not mutually exclusive. In general terms we may consider the *bullae* as convenient containers or envelopes to isolate the tokens representing a particular transaction. The *bullae* fulfilled the same function as the early tablets. They were economic records. The former used a three dimensional system of tokens in the shape of cones, spheres, etc., enclosed *in* a clay ball; the latter used two dimensional written signs, some in the shapes of cones, spheres, etc., written *on* a clay ball. The *bullae* could therefore be viewed as the prototypes of the tablets explaining the choice of clay for writing and the convex shape of the tablets.

3. The First Impressed Signs

The *bullae* provided the great advantage of securing the tokens tightly and presented a surface where seals could be impressed for authentication. Their disadvantage was to totally hide the tokens, thus any verification necessitated the breaking of the *bullae.* To overcome this difficulty some *bullae* bear on their outer surface impressed marks repeating the number and the shape of the tokens inside: circular impressions for discs and spheres, conical impressions for cones (plate XIII, ill. 10). The innovation was ,of course, of great convenience as it allowed one to "read" at all times the amount and kind of tokens without breaking the *bulla.* These marks may be viewed as the crucial link between an archaic system of recording in three dimensions and writing. As soon as the system of marks on the exterior of the *bullae* was generally adopted and understood, it obviously made the system of tokens inside the *bulla* superfluous and obsolete and tablets with impressed signs made their appearance (plate IX, ill. 5).

4. The First Incised Signs

The signs impressed with the tokens themselves, with the thumb, or with the blunt tip of a stick did not allow much precision. Using such an instrument, it was not possible to differentiate between the

[11] Pierre Amiet. "Il y a 5000 ans les Elamites inventaient l'ecriture, *Archeologia*, No. 12, 1966. pp. 20-22.
[12] Pierre Amiet. *Elam.* Auvers sur Oise, 1966. pp. 70-71.

representation of a sphere or a disc, between a cone and a tetrahedron. Signs incised with a specialized pointed tool or stylus were developed which could render more complicated shapes with accuracy. Though these signs were pictographs, they did not represent the shapes of the items themselves but rather the tokens of the previous recording system. As writing was established, new vocabulary expressing new inventions such as metal tools, the plough, chariot, etc., consisted of true pictographs, drawn according to the objects themselves.

Conclusions

Writing may not have been, as previously assumed, brought by such newcomers to Mesopotamia as the Sumerians, but may represent a new step in the evolution of a sophisticated system of recording which was indigenous to the Middle East since the IXth Mill B.C. This accounting system was based on tokens of various geometric and odd shapes. It was used without apparent modifications until the IVth Mill B.C., when what should have remained a marginal invention, the use of a clay envelope to hold the tokens of a particular transaction and bear the relevant seals, was to usher the system in a drastically new course. Check marks noted on the surface of the *bullae* to repeat for convenience the number of tokens inside proved effective and soon supplanted the old system of tokens. If this hypothesis is correct it would shed light on some of the characteristic features of the archaic tablets and in particular their material and convex forms. It would explain the instant standardization of writing in a widespread area as well as the abstract shapes of all the numerical signs and of a score of commodities of daily use. This theory alters our previous conception of human cultural development,

First, it pushes back by five millennia the origins of writing. Second, it evidences the existence of a sophisticated system of recording prior to writing. Third, it suggests the presence of a *lingua franca* in the form of tokens which was shared by a number of prehistoric and early historic cultures of the Middle East and may have extended as far as Nubia and the Indus Valley.

Of course, it also raises many more questions, among which some of the more crucial would be what was the prototype of the remaining some 800 signs of the Warka tablets? Which culture of the Middle East from Anatolia to the Zagros invented the token system? How did it become so widespread before any trade is evidenced? Did tokens of identical shapes bear the same meaning in Jarmo and Khartoum? Although they may remain unanswered, these questions are certainly exciting challenges. They will hopefully create a new awareness among excavators and will attract attention to the small, but maybe significant, tokens which up to now have been labeled in site reports as "objects of uncertain purposes."

Der Meister sol auch kennen schreiben und lesen: Writings about Technology ca. 1400-ca. 1600 A.D. and their Cultural Implications

by

Bert S. Hall

In the preface to his now-famous *Medieval Technology and Social Change* Lynn White, jr. comments on the illusory quality of much conventional history. The historian shapes his craft on the implicit belief that "surviving written records provide us with a reasonably accurate facsimile of past human activity."[1] Until quite recently, however, much of the human race lived out its existence in illiteracy or in only marginal literacy, and thus in a sort of sub-history screened from the historian's gaze. The historical study of technology is particularly affected by this condition, for technology was chiefly the affair of subliterate groups in antiquity and through most of the middle ages. As White and others, both historians and archeologists, have shown, it is possible to explore the history of technology despite the paucity of written records, and in fact, the results of their investigations have been extremely fruitful. On the basis of this work, it is now possible to pose a different type of question than would have been possible a decade or so ago. The purpose of this paper is to explore the boundary between the sub-literate, and therefore sub-historical, phase of technological development and the beginnings of a continuous, cohesive written literature concerning technological matters. It is by no means an exhaustive study of this topic, but rather an attempt to pose the issues in ways that will permit further investigation.

Given that technology remained for many centuries after the appearance of writing mainly the concern of sub-literate groups, at what point do we begin to find technical writings appearing in significant numbers? What explanations can we advance for the change from orality to literacy? And what difference did the change make to technology itself and to the societies that encouraged technological effort? In all pre-literate societies, and among those sub-literate groups who practiced certain crafts in antiquity and the middle ages, pertinent technical information was obviously recorded and transmitted through memory and via oral and manual teaching. That this was an effective and flexible manner of proceeding is manifest to anyone who has ever studied, let us say, a medieval cathedral. Only in the two hundred year period from roughly 1400-1600 A.D. did this traditional pattern of information storage and retrieval begin to be supplemented on a large scale by written documents, e.g. handbooks, notebooks, and treatises, and by drawings or sketches intended for preservation. The tradition of putting pen to paper for these purposes had barely begun when the printing press was invented, and after a certain period of delay, the press became in turn an important means of distributing technological information to a wider public. Technology went through two shifts of medium in the period under consideration, from oral to written and from script to print. (It should be noted that neither shift entirely displaced the former media.) The experience of working in a new medium posed certain problems for technical writers, e.g. how to express complex ideas in words and pictures, or how to retain proprietary rights to new ideas. None of these was solved completely, but the way technicians dealt with them is of interest.

1 (Oxford and New York: Oxford University Press, 1962), v.

The appearance of the illustrated treatise and its printed descendant fostered contacts between the technician's work and the world of high culture, since the treatises were apparently read not only by craftsmen, but by aristocrats, lawyers, and intellectuals as well. The presentation of technological matters before such an audience helped heighten the technician's sense of self-consciousness and make him more willing to justify his activities in ways acceptable to Renaissance high culture.

The Shifts of Medium

The first significant change in the traditional way of remembering and teaching technical data begins to take form only about 1400 A.D. with the appearance of large numbers of manuscripts dealing with technology. To be sure, there are treatises on technical matters from before this date, but if one surveys the earlier writings in a general way, one cannot help but be impressed by the very small number of such works, by the way they are so widely separated in time and space, and by the fact that their authors were seldom practicing craftsmen. The rare exceptions such as Vitruvius[2] or the Alexandrian technicians from antiquity,[3] or Theophilus Presbyter[4] or (possibly) Villard de Honnecourt[5] from the middle ages, are precious to the interested historian precisely because he so seldom encounters their like. Illiterate craftsmen had neither the need for such treatises nor the means to produce them; the literate lacked the knowledge necessary to write on technical subjects and the interests necessary to create an audience for such works, were they to be written.[6] Indeed, even the skills necessary to copy correctly an older work were often beyond the scope of technologically untutored scribes, as is amply revealed by the present state of the *Mappae clavicula*, a collection of workshop recipes badly garbled by several centuries of transmission through monastic scriptoria.[7] The world of the learned and that of the craftsman were so widely separated as to preclude any serious possibility of creating a continuous, cohesive body of detailed technical literature, either with respect to initial authorship or the subsequent labor of copying necessary to preserve any manuscript book. The exceptions mentioned above merely prove that the separation was not absolute, and like a mountain rising from a plain, call attention to the absence of conspicuous features in the rest of the landscape.

After 1400, by contrast, there is a veritable flood of technical treatises; the written record becomes so rich that the historian is almost innundated with documents. There is a high degree of continuity between individual examples, indicating that full or partial copying was frequent and usually intelligently carried out. Not all the authors of such works are known to us (in fact there is a serious problem about what constitutes authorship within these groups of texts), but when we can discover an author's or copyist's name, it often reveals a practicing technical specialist. The quality of the anonymous texts is likewise sufficiently high to suggest "professional" technicians as authors. The change in quantity of textual material is too great and the change takes place too abruptly for statistical explanations based on the survival rates of perishable manuscripts to work in this case.[8] The quantitative change as well as the qualitative character of the surviving works argues quite strongly that men of the fifteenth century rather suddenly began to resort to pen and paper in a field where their fathers had been satisfied with memory and the spoken word. We shall discuss below what seems to underlie this new impulse.

[2]F. Granger (ed. and trans.), *Vitruvius on Architecture* (London: Heinemann, 1934).

[3]A. G. Drachmann, *Ktesibios, Philon, Heron* (Copenhagen: 1948); *The Mechanical Technology of Greek and Roman Antiquity* (Madison: University of Wisconsin Press, 1963). Eric W. Marsden, *Greek and Roman Artillery* (2 vols.; Oxford: Clarendon Press, 1969-71) gives detailed information on the military technology of these men.

[4]C. R. Dodwell (ed. and trans.), Theophilus Presbyter, *De diversis artibus* (London: Nelson, 1961); John G. Hawthorne and C. S. Smith (eds. and trans.), *On Divers Arts* (Chicago: University of Chicago Press, 1963).

[5]Theodore Bowie (ed.), Villard de Honnecourt, *Sketchbook* (2nd ed.; New York: Wittenborn, 1962). Cf. Alfred Darcel (ed.), *Album de Villard de Honnecourt* (Paris: Laget, 1968).

[6]Lynn White, jr., "Medieval Engineering and the Sociology of Knowledge," *Pacific Historical Review*, 44 (1975), 1-21 discusses several aspects of this topic.

[7]C. S. Smith and John G. Hawthorne, *Mappae clavicula: A Little Key to the World of Medieval Techniques* ("Transactions of the American Philosophical Society, N.S. *64*, pt. 4;" Philadelphia: 1974), 14 ff.

[8]Most of these remarks are discussed in greater detail in my dissertation, "The Manuscript of the So-Called 'Hussite Wars' Engineer' and Its Technological Milieu: A Study and Edition of *Codex latinus monacensis* 197, Part 1," (University of California, Los Angeles: Department of History, 1971). This work is scheduled to appear in 1978 as *The Technological Illustrations of the So-Called "Anonymous of the Hussite Wars,"* (Wiesbaden: Dr. Ludwig Reichert Verlag).

The second shift of medium saw the augmentation of the written tradition by the printed book. We are so accustomed to taking for granted transferences of this sort that we should pause to think that, in this case, it was far from easy. Technicians had only recently begun to write about their crafts, and much of the early literature was crude and inelegant in the extreme. The highly illustrated character of many of the early manuscripts would have made them difficult or prohibitively expensive to publish through the press. There must have been serious questions about the audience for any work of this type, since in comparison with such staples of the trade as theology or *belles lettres* technical works would seem to have had limited sales appeal. Quite probably, even if writers were available to do works on technology for the press, printers might well have been reluctant to undertake anything of the sort. Significantly, only a few of the fifteenth century manuscript works reached print, and these under unusual circumstances; most of the printed works appeared only in the sixteenth century, and they were deliberately composed with the press in mind. From the 1530's onward there is a slow but steady increase in the number of printed books on technology, and by the latter third of the century, print had become a suitable medium both for descriptive works and for presentation of technical innovations, as well as discussions of the significance of technology in general. This movement toward a fully public medium completed the transformation initiated with the shift toward manuscripts in the fifteenth century. Technology, which had once been nearly the exclusive possession of a particular craft group,[9] was now being displayed and discussed before a wide audience of educated, but unspecialized readers. By 1600, technical information was being transmitted through three main channels: the older, oral medium of the craft group; the manuscript which circulated among interested parties; and the printed work intended for a "semi-popular," moderately learned, lay public. With minor modifications, this situation persisted until the late eighteenth century and the rise of schools of engineering. Its final transformation came only in the following century with the rise of industrial societies and the beginnings of mass secondary education.

The Manuscript and Printed Works

The body of books, treatises, manuals, pamphlets, and notebooks in which these trends reveal themselves is far too large to be described completely in this paper. Indeed, there is still insufficient scholarship in this field to permit a complete description under any circumstances. Some brief remarks are necessary, however. The manuscript works of the fifteenth century appear first in Italy and those parts of central and eastern Europe where German was the vernacular. I have argued elsewhere that these regional differences yield two distinct schools or traditions which have only modest mutual influence before the end of the century.[10] The trans-Alpine school begins with Conrad Kyeser's *Bellifortis* (ca. 1405), an illustrated treatise with metrical Latin text that actually deals with far more than its title, "strong war," would indicate.[11] Kyeser's book spawned a host of copies, imitations and adaptations; parts of it were still resonant in sixteenth or seventeenth century treatises on the arts of warfare. An apparently independent progenitor of later treatises was the very popular *Feuerwerkbuch* which must date from around 1420. *Feuerwerkbuch* is a practical, didactic manual for artillerymen and consists principally of recipes for pyrotechnic compounds.[12] Many German works contain portions of the two ancestral books, and virtually all later technical writings owe something to either by way of

[9]The use of the term "craft group" rather than "guild" is a deliberate attempt to avoid some of the institutional complexities associated with the latter. Not all medieval craftsmen were organized into guilds, nor were all guilds composed of craftsmen. Still, it is plausible to assume that changes in guild structure and membership may have had some influence on the movement toward writing, and conversely, the existence of a written tradition may have influenced the history of guilds. In any case, the study of the relationship between guilds and the documents under consideration would necessarily require more work than is presently available, and it cannot be dealt with in this paper.

[10]See note 8. The most extensive bibliography of such works is still found in Max Jähns, *Geschichte der Kriegswissenschaften, vornehmlich in Deutschland*, Vol. 1 (München and Leipzig: Oldenbourg, 1889). This should be used with caution, however.

[11]Götz Quarg (ed. and trans.), Conrad Kyeser aus Eichstätt, *Bellifortis* (2 vols.; Düsseldorf: V.D.I. Verlag, 1967). Cf. the critical review of this work by Hermann Heimpel in *Göttingische gelehrte Anzeigen, 223*, Heft 1/2 (1971), 115-48.

[12]The only available edition is that of Wilhelm Hassenstein, *Das Feuerwerkbuch von 1420* (München: 1941). Hassenstein was a dedicated *Oberregierungsrat* in the NS regime, and his work should be used with extreme caution. This edition is virtually impossible to obtain in Europe or North America, but I would be happy to make xerox copies available to any interested scholars for research purposes.

inspiration. An exception is the small group of mason's treatises which appeared in the decades around the turn of the century.[13] All of these works in turn stand as forerunners of such better-known treatises as Albrecht Dürer's on fortifications.[14]

The Italian school begins with Giovanni Fontana's small *Bellicorum instrumentorum liber* (ca. 1420),[15] but it acquires its lasting character only after the period of Brunelleschi's activity in Florence. Mariano di Jacopo, called "Il Taccola," a Sienese friend of Brunelleschi, left two treatises (one unfinished) from the 1430's and 1440's.[16] He was followed by another Sienese, Francesco di Giorgio Martini, whose compendious writings and illustrations on art, architecture and military engineering continue to be repeated even in seventeenth-century treatises.[17] Among those who studied Francesco's works was the young Leonardo da Vinci. Leonardo's notes and sketches are, of course, *sui generis,* but their mechanical and architectural portions can clearly be located within the stream of sub-Alpine fifteenth-century treatises.[18] We now know that Leonardo did rework many rough notes and sketches into a more formal treatise on theoretical and applied mechanics which he hoped to publish. It has survived as the recently-rediscovered *Codex Madrid* I.[19]

The movement toward printed technological works can be dated formally from 1472, when Roberto Valturio's *De re militari* was published.[20] It was followed in Germany in the 1480's by Lorenz Lechler's little didactic treatise on building in the Gothic manner.[21] Despite these early harbingers, only after 1529 does the printed tradition gain momentum. In that year *Feuerwerkbuch*, after a long history in manuscript, was printed in two unrelated first editions.[22] (These three works were to my knowledge the only ones to make an intact transit from script to print.) No simple scheme can encompass all the sixteenth century printed works on technology. There is a strong military stream stemming from *Bellifortis* and *Feuerwerkbuch* and extending unbroken into the seventeenth century; unfortunately, almost no scholarship exists on those works. There is a group of descriptive books more familiar to historians and concerned with mining and metallurgy: Vanoccio Biringuccio's *Pirotechnia* (1540),[23] Georgius Agricola's *De re metallica* (1556),[24] and Lazarus Ercker's *Beschreibung Allerfürnehmisten Mineralischen Ertzt und Berckwercksarten* (1574).[25] The most interesting groups of books for our purposes bears the generic name "Theaters of Machines." These highly illustrated Baroque encyclopedic treatises sought to present, often with more vigor than rigor, unusual or novel machines to the general educated public.[26] The earliest of the "Theaters," and the source of their generic name, was Jacques Besson's *Theatrum instrumentorum et machinarum* (ca. 1569-70). Besson's book went through at least eleven

13Lon R. Shelby, *Gothic Design Techniques: The Fifteenth-Century Design Booklets of Mathes Roriczer and Hanns Schmuttermayer* (Carbondale, Ill.: Southern Illinois University Press, 1977). My thanks to Professor Shelby for allowing me to read a copy of this work in proof sheets.

14Alvin E. Jaeggli (ed. and trans.), Albrecht Dürer, *Unterricht über die Befestigung der Städte, Schlösser und Flecken* (Dietikon: Stocker, 1971).

15München, Bayerische Staatsbibliothek, *Cod. icon.* 242; no modern edition has been made. See Marshall Clagett, "The Life and Works of Giovanni Fontana," *Annali dell'Istituto e Museo di Storia della Scienza di Firenze, 1* (1976), 5-28.

16Gustina Scaglia (ed. and trans.), Mariano Taccola, *De machinis: The Engineering Treatise of 1449* (2 vols.; Wiesbaden: Dr. Ludwig Reichert Verlag, 1971). See also Frank D. Prager and Gustina Scaglia, *Mariano Taccola and His Book* De ingeneis (Cambridge, Mass.: MIT Press, 1972).

17Corrado Maltese (ed.), Francesco di Giorgio Martini, *Trattati di architettura, ingegneria e arte militare* (Milan: Polifilio, 1967).

18Bertrand Gille, *Engineers of the Renaissance* (Cambridge, Mass.: MIT Press, 1966). Gille may have overreacted to the tendency to celebrate Leonardo's accomplishments, but his work still contains one of the best studies of the relationship between Leonardo and the evolving tradition of fifteenth century writings on technology.

19Ladislao Reti (ed. and trans.), Leonardo da Vinci, *The Madrid Codices* (5 vols.; New York: McGraw-Hill, 1974). Cf. Bert S. Hall, "The New Leonardo," *Isis, 67* (1976), 463-476.

20Erla Rodaliewicz, "The *editio princeps* of Roberto Valturio's *De re militari* in Relation to the Dresden and Munich Manuscripts," *Maso Finiguerra, 5* (1940), 15-82.

21See note 13.

22(Augsburg: Heinrich Stainer, 1529) and (Strassburg: Christain Engenolphen, 1529), the latter under the title *Büchsenmeisterei: von Geschoss, Büchsen, Pulver, Salpeter, und Feuerwerk.*

23Cyril Stanley Smith and Martha Teach Gnudi (eds. and trans.), *The Pirotechnia of Vannoccio Biringuccio* (2nd ed.; New York: Basic Books, 1959).

24Herbert Clark Hoover and Lou Henry Hoover (eds. and trans.), (London: The Mining Magazine, 1912).

25Anneliese G. Sisco and C. S. Smith (eds. and trans.), Lazarus Ercker, *Treatise on Ores and Assaying* (Chicago: University of Chicago Press, 1951).

26No general work on the "Theaters" has been done. A semi-popular discussion with a generous selection of representative sample plates is found in Alexander Keller, *A Theater of Machines* (New York: MacMillan, 1965).

editions by 1602 and was translated from its original Latin into French, Spanish, Italian and German.[27] Its influence can still be seen as late as the great *Encyclopedie* of the French enlightenment. For the sake of simplicity we shall ignore all Besson's numerous followers in the genre save one. In 1588 Agostino Ramelli, an engineering officer to the crown of France, published a lavish French and Italian "Theater" under the title *Le diverse et artificiose machine.* Though it lacked the number of editions Besson's enjoyed, Ramelli's book in both its original and a later German edition was quite influential.[28]

The Authors

To understand what took place between Kyeser and Ramelli, we need first to look at the authors of technological works. The earliest treatises, Kyeser's *Bellifortis* and Fontana's *Bellicorum instrumentorum liber* were written by men who were trained as physicians and astrologers. This combination, so strange to the modern mind, was of course quite normal in Europe since the revival of astrology in the thirteenth century. No one could hope to affect a successful cure of any disease if he remained ignorant of the complex of causal connections between his patient, the disease, and the astral and planetary influences. This in turn led medical astrologers to interest themselves in instruments, and especially in clocks. In a recent study, Lynn White, jr. has pointed out how indebted we are to the medical stargazers for information about late medieval technology;[29] we might wish to mention in passing only Guido da Vigevano, an Italian physician to the French court, whose little treatise of 1335 on the technological needs of a crusading army has recently been edited by A. R. Hall.[30] In their roles as physicians-astrologers commenting on technology, Kyeser and Fontana represent an older tradition. It would reappear from time to time (Agricola was a physician by profession), but most of the works that followed were composed by men of different professional character.

In Germany and in Italy in the fifteenth century, the needs of warfare were changing rapidly. The fact that Kyeser and Fontana devoted parts of their work to military matters should not surprise us; indeed, their interests provide an important clue to the shift toward the literate medium. Late medieval warfare is still a subject of some confusion, but it is probably safe to generalize by saying that war was becoming both more complex and more professionalized around 1400. The older ways of the feudal levy and the simple siege train of a few towers and catapults were being challenged by a host of hellish new machines led by firearms. Indeed, one reason for our confusion about this period's military history is our inability to assess the rapid changes in machines and their collective impact. Technological authors, on the other hand, seem less uncertain; many of them act as if war were primarily a matter of pitting one machine against another. They were probably correct to this extent: as war became a more complex tactical and technological business, the men who knew the machines of war, especially firearms, tended to become more important and valued servants of the political rulers.[31] This helps explain why, in Germany at least, there is a pronounced shift in the generations after Kyeser toward professional authorship of technological texts, and the profession most often represented is that of the *Büchsenmeister* or master gunner.

I have not been able to determine the full range of the *Büchsenmeister's* normal duties in fifteenth-century Germany, but he seems to have been responsible in general ways for the making and repair of cannon and other siege engines, gunpowder production and transport, fortifications (particularly anti-

[27] The original edition appeared without indication of date or publisher under the title *Instrumentorum et machinarum: liber primus.* On internal evidence this must have appeared after June 27, 1569. There were four editions in 1578 under the standard title in Latin, French and both languages, all published by Bartholemy Vincent in Lyons. Vincent also produced another French edition (1579), another Latin edition (1582), and an Italian edition (1582). A French version was published in Geneva by I. Chouët in 1594, a German version by J. Folliet in Mümbelgart in 1594, and a Spanish edition by H. Cardon in Leon in 1602.

[28] Eugene S. Ferguson and Martha Teach Gnudi (eds. and trans.), *The Various and Ingenious Machines of Captain Agostino Ramelli* (Baltimore: Johns Hopkins University Press, 1976).

[29] Lynn White, jr., "Medical Astrologers and Late Medieval Technology," *Viator*, 6 (1975), 295-309.

[30] A. Rupert Hall, "Guido's *Texaurus*, 1335," in Bert S. Hall and Delno C. West (eds.), *On Pre-Modern Technology and Science* ("U.C.L.A. Center for Medieval and Renaissance Studies, *Humana Civilitas*, Vol. 1;" Malibu, Calif.: Undena, 1976), 11-52.

[31] Evidence for this claim is admittedly sketchy, largely because the lives of early technicians still represent *terra incognita* in all but a few cases. See Lynn White, jr., "Jacopo Aconcio as an Engineer," *American Historical Review, 72* (1967), 425-444; also Carlo Promis, *Biografie di ingegnieri militari italiani dal secolo xiv alla metá xviii* (Turin: 1874).

firearms fortifications), and the use of all these in wartime, and general peacetime duties such as roadwork and public fountains.[32] It would seem as if the old *ingeniator,* the "master of engines" of the thirteenth century was becoming a more general technologist, an "engineer" in something approaching the modern sense. If this is true, and if the assumption of increased value laid on such services is true as well, then the role of such men in putting pen to paper and recording their craft is more easily understood. One passage in the *Feuerwerkbuch* hints at this connection. Describing the duties of a good master gunner, the anonymous author mentions the conventional pieties, honoring God and one's prince, but he then goes on to insist that the master must also be literate, *der Meister sol auch kennen schreiben und lesen.* Why? Because without literacy he could not possibly "keep in mind all the elements that belong to this art."[33] We should probably not assume from this that technological complexity alone necessitated literacy, but rather that such complexity, combined with rising literacy rates, combined with enhanced social and professional status, all worked together to create a class of literate technicians who could and did speak for themselves, taking over the roles once held by physician-astrologers or other interested amateurs of technology.

The Germans move rapidly down the road toward almost complete militarization of the technical literature by the sixteenth century. The Italian school moves in slightly different directions toward a blend of artistic and engineering interests (to which we shall return shortly), and retaining an "amateurish" quality in its writings for some length of time. In Italy there is certainly no lack of interest in matters military (recall that Valturio's *De re militari* was so popular that it was the first to reach print), but the art of war is mixed in with and leavened by a broader set of interests in what we might call "civil" technology. Several studies by Frank Prager and Gustina Scaglia have pointed to the construction of the cathedral in Florence and to Brunelleschi in particular as the font of this concern with construction technology and its affiliated arts.[34] Brunelleschi has left us no direct notes or drawings, but Taccola apparently copied and recorded many of his ideas, whence they spread widely into the literature, to be received by, among others, Leonardo da Vinci. Taccola was himself a notary by trade, though he moved in Brunelleschi's circle. Taccola's Sienese follower, Francesco di Giorgio Martini, earned his living primarily as an architect but he also did scupture as well as writing technical treatises. Leonardo is so broad in his interests as to resist easy categorization, but it is worth recalling that for most of his life after 1482 he earned his living as something like a "consulting engineer" and that his duties in this role included military technology and hydraulic engineering. The Italians seem to have begun writing about technology for many of the same reasons as the Germans. The differences between the patterns of such writings in the two countries is probably due to the different social and intellectual environments. We have learned to expect breadth in all things from men of the *Quattrocento*, and it is not surprising to find it in technology as well. By the 1500's, however, signs of specialization appear even in the Italian treatises. Biringuccio, for example, rarely wanders beyond the limits of his craft, broad though these were. Ramelli retains some of the older impulse to wander widely, but no reader of his work would have much difficulty guessing that he was a military engineer by profession. In short, despite initial differences, there is an increasing tendency toward specialization in technological writings, regardless of the nationality of their authors. More than anything else, this shows the increasing maturation and formalization of the movement toward literacy born among the fifteenth-century technicians.

Art and Technology

The blending of artistic and technological interests mentioned above represents another of technology's cultural relationships developed in this period. On the social level, the interest of artists such as Leonardo or Dürer in technology can be explained by reference to their relatively undifferentiated social

[32] Leonardo's surviving draft of a letter to Ludovico Sforza, "Il Moro," seeking employment as an engineer and describing his many skills is perhaps the best concise statement of what an engineer was expected to do in this period. Gille, *Engineers*, 125-126 quotes it in full. One might object that Leonardo was a polymath, and in any case everyone exaggerates his range in job-seeking letters, but there is ample evidence in other treatises and notebooks that Leonardo's list is essentially valid for his less gifted contemporaries. The German *Büchsenmeister* was probably expected to know about as much as his Italian counterpart.

[33] Cited in Hall (see note 8).

[34] Frank D. Prager and Gustina Scaglia, *Brunelleschi: Studies of His Technology and Inventions* (Cambridge, Mass.: MIT Press, 1970); cf. Prager and Scaglia, *Taccola*.

milieu; artists and technicians were both regarded as craftsmen in the fifteenth century (though they were more separated in the sixteenth). There are deeper connections, however, that need some exploration. When I first began to examine the documents I am discussing, I noticed that the more interested a particular text was in mechanics or architecture, the more likely it was to be profusely illustrated. As I studied them further, I realized that I was approaching them with the wrong presuppositions. I did not have in hand a group of illustrated texts, but rather a group of pictures with running commentaries. We scholars are men and women of the book; to us it is axiomatic that words convey primary meanings, while pictures are secondary, or even purely ornamental. But consider the difficulties of a writer, let us say, on applied mechanics. It is quite difficult even today to produce a verbal description of a machine, or building, or complex technical process that can convey more than superficial meanings. Yet all modern languages used in industrialized societies are extremely rich in technical terms with exactly defined meanings, and all users of these terms are likely to have had extensive training in precisely how to employ them. The corresponding difficulties of a fifteenth-century writer have to be imagined. Though some technical terminology was available—most notably in such tightly-knit fields as masonry—most verbal descriptions had to be fashioned from languages still poorly equipped for the business of technical discourse. (There is, by the way, a serious lack of philological scholarship on questions of technological vocabulary formation in the pre-industrial world.) If neither Latin nor any of the vernaculars was really capable of precise descriptions in this realm, then pictures had to be used for primary meanings, while words were secondary, or even entirely unneccessary. Many of the "texts" (and that word is now to be understood as having quotation marks) from Kyeser to Ramelli, including Leonardo, are very nearly "picture-books" of technology with verbal comments. At times, the primacy of the picture is made unmistakably clear, as for example in a German work of the 1430's which dispenses with text altogether.[35] Lest it be thought that such books were for subliterates only, let me point out that the work in question was a presentation copy for the Emperor Sigismund, to whom it is dedicated (naturally by means of a picture; plate XV, ill. 1).

This emphasis on the pictorial has a number of implications. First, it forced an alliance between art (understood in its commonplace sense of graphic representation) and technology; anyone interested in compiling a work on mechanics had to be an artist himself to some degree, or had to work closely with artists, as we know Kyeser did.[36] This would make it possible to do an art-historical analysis of such works as they reflect changing styles of representation. More to the point, it raises the question of how changes in art affect the technologist's ability to conceive and express his subject matter. As Flemish realism and Italian linear perspective both penetrate the world of technical writings, they displace the old medieval illuminator's style of representation with a pictorial language that is similar to our own. Thus the later texts look "better" to the modern eye. Compare, for example, *Das mittelalterliche Hausbuch* (c. 1480)[37] or Dürer's treatise on fortifications with *Bellifortis* or any of the early German works (see plates XVI and XVII, ills. 2 and 3). It would be unfair to suggest comparing Leonardo with Taccola or Fontana, but the point would be the same. Within the realm of "pure art" we have learned to be suspicious of words like "better." "Is Raphael a 'better' painter than Picasso?" sounds like a silly question to us. Yet when seen from another perspective, that is, the value of pictures as means of conveying information, the question of styles is somewhat different. The later texts are "better" visually because their creators, having learned the Flemish tricks and Italian rules, are better able to treat the complex relationships of mechanical parts in three-dimensional space, and to portray them with less error or distortion. Thus the inventions of the artist provided the technologist with a powerful new vocabulary for portraying and transmitting information about his subjects.

It is also possible that art and changes in artistic modes of representation may have had some influence on the very process of invention itself. We have only a dim understanding of how mechanical inventions

[35] Vienna, Waffensammlung des kunsthistorischen Museums, MS P 5014, frontispiece. The relationship of words and pictures in a number of late medieval works needs further consideration. It is probably incorrect to assume that picture-Bibles, for example, were entirely considered as suitable only for the poor or illiterate. Conversely, texts with illustrations often treat the picture as pure ornament without any intrinsic descriptive value whatsoever. Whatever late medieval men thought about pictures in relationship to words, their attitudes were surely very different from ours.

[36] Lynn White, Jr., "Kyeser's 'Bellifortis': The First Technological Treatise of the Fifteenth Century," *Technology and Culture 10* (1969), 437.

[37] Helmuth Th. Bossert and Willy F. Storck (eds.), *Das mittelalterliche Hausbuch* (Leipzig: Seeman, 1912).

arose in the pre-modern era, but we are certain that they were not, as in our own day, logically conceived applications of previously thought-out scientific principles. If science played no strong role, what did? One answer may be found in Leonardo da Vinci's works, parts of which seem to display a series of imaginative variations on the morphology of machine parts as the intermediate steps toward new mechanical designs.[38] Leonardo, in a word, seems to have thought over his creations on paper through use of visual imagination (plates XVIII and XIX, ills. 4 and 5). It would be risky to extrapolate from Leonardo to his less gifted contemporaries, but there is nevertheless some reason to believe that other technical authors did approximate his method of inventing "on paper". To some degree, therefore, the ability to sketch with speed and accuracy may have been a means of conceiving new ideas as well as portraying existing devices. At the very least, we should expect artistically trained technicians to begin to validate their visions of machines by sketching them on paper as a preliminary step. The drafting pen may well have become a tool—albeit an imperfect one—to aid an inventor's intuition during this period.

By the sixteenth century another artistic device was coming to have an impact on technological writings, the copperplate engraving. It was possible in the fifteenth century to print a work requiring some visual component, e.g. Valturio, by means of woodcuts, but the development of printing from engraved metal plates, which did not become common until the 1500's, was an enormous boon to technical authors. Their visual ideas could be reproduced with great fidelity in almost infinite quantities via copperplates. The great degree of dependence of authors like Besson and Ramelli on this particular artistic device is manifest at first glance (plate XX, ill. 6). Quite possibly the relatively late appearance of such works, that is the timing of the shift to print, was in part a function of the spread of cheap copperplate engraving techniques. Moreover, since Besson and Ramelli, unlike their fifteenth century manuscript counterparts, could not themselves produce their drawings in the form in which they would circulate, there had to develop a complex network of connections between author and engraver, financier and printer before such works could be published. This too may have had a retarding influence on the appearance of such works.

Authors and Audiences

Another aspect of the culture of technology emerges if we ask about the way such works were used by their audiences. Obviously no one writes a book without some sort of user in mind (except, of course, for such purely private jottings as we see in Leonardo's *Notebooks*). It is difficult for us to place ourselves imaginatively in the milieu of the authors and to see just exactly who they had in mind when composing their works. At times the question is answered for us. *Feuerwerkbuch*, for example, is obviously by a gunner for the use of other gunners. Most of its contents have nothing to do with mechanics or ballistics as a science, but instead seem to have served as a sort of cookbook for the gunner, and some of the copies I have examined show stains and charring to indicate that they were used in just such a fashion in the workshop or arsenal.

The question becomes more involved when we turn to other works. The German tradition stemming from Kyeser seems to have almost no respect for what we consider authorship or textual integrity. Kyeser himself is, next to Leonardo, the most idiosyncratic and self-possessed of authors. His work is shot through with his personal likes and dislikes, leaning heavily toward the latter, and he even offers us his portrait and self-composed epitaph. (His is the first realistic portrait of an author since antiquity, by the way.) His followers, however, were of a different stripe. They tended to decompose, recompose, alter, edit, and delete old texts and then mix the result with new material from available sources. This chaotic blend of traditional and novel contents is the absolute despair of the orderly-minded historian, who wants his "texts" in neat rows suitable for making charts in introductions, but the Germans at least preferred a kaleidoscopic approach to such matters. We do find an occasional name attached to a work,

[38] I have analyzed one such case in detail. See Bert S. Hall and Ian Bates, "Leonardo, the Chiaravalle Clock and Epicyclic Gearing: A Reply to Antonio Simoni," *Antiquarian Horology* 9 (1976), 910-917. The same general method seems to have been used in Leonardo's invention of a pendulum regulator for clockwork; cf. the review cited in note 19 above. A comparable, if considerably simpler case concerns the invention of a parachute in the 1470's. See Lynn White, jr., "The Invention of the Parachute," *Technology and Culture*, 9 (1968) 462-67.

usually a *Büchsenmeister*'s, but it merely refers to the compiler, not the author in most cases. The epitome of the compilation approach is reached when we find deluxe encyclopedias of technology compiled from many sources and obviously made under noble patronage by highly skilled illuminators to grace the palace library;[39] in fact one can sometimes trace the same drawing as it climbs the social scale from crude original to final entombment in a deluxe copy within two or three generations. The fact that aristocrats were interested enough in technological writings to purchase manuscript copies or to patronize "improved" editions is an index of the rising social status of technical authors and their subject matter. We should also note that the deluxe copies, like the "picture book" presented to Emperor Sigismund, seem slightly biased toward military technology; this suggests that a certain selectivity in what was patronized may have contributed to the military character of the later German tradition. In any event, the German evidence shows what we might call a "semi-public" literature of technology in which texts, drawings, and ideas are considered common property. They are circulated among interested parties and are available to anyone who wishes to copy or use them, without any apparent legal or social inhibitions.

The Italians seem to have more definite ideas about such matters from the very start. Brunelleschi, the man responsible for many of the new ideas in the Italian tradition, is reported by Taccola to have cautioned him against making such knowledge too widely available.[40] Fear of theft, not deliberate obscurantism, was Brunelleschi's motive. What his complaints seem to point to is exactly the sort of attitude displayed in the German tradition. It was, of course, an attitude that was completely functional in the pre-literate days of technological development, and it would persist for some time into the sixteenth century. Taccola ignored Brunelleschi's advice at his peril and was repeatedly plagiarized, as was Francesco di Giorgio Martini, who complained bitterly about such treatment.[41] We no longer like to explain Leonardo de Vinci's mirror writing in the *Notebooks* as a response to fear of persecution (it came naturally to him as a left-handed person), but when we find him employing it in a draft of what was to be a formal treatise on mechanics—*Codex Madrid* I—we might want to attribute to him some desire to maintain proprietary rights in his ideas by making casual copying difficult. Elsewhere in his works Leonardo seems to fear being copied without credit.

What all these men seem to want is something like modern copyright or patent protection. The history of legal devices to accomplish such ends is long and very complex.[42] Feudal rulers could assign various rights or privileges to certain of their subjects through "patentes" or letters of permission. Municipal councils could act in similar fashion with respect to items of manufacture or commerce which were of interest to the bourgeoisie of late medieval Europe. Inventors seem to have had mixed success in obtaining the type of protection granted to viscounts, monasteries, or wool-merchants. Brunelleschi was able, for example, to negotiate from the Florentine *signoria* a form of monopoly right to a scheme he devised for transporting building materials to the city for its cathedral.[43] Johann Gutenberg, on the other hand, died in poverty when he was unable to protect his invention against expropriation by others. The printed book—if not the press that produced it—was subject to a degree of protection because it represented a salable item in which the printer might have considerable capital investment. Protection

[39] Among the best of these deluxe editions is Erlangen, Universitätsbibliothek, MS 1390, known as the *Kriegsbuch* of Ludwig von Eybe zum Hartenstein (der Jüngere) and dating from ca. 1500. Another splendid specimen is Weimar, Nationale Forschungs- und Gedenkstätten der klassischen Deutschen Literatur, MS fol. 328, the so-called "Ingenieur Kunst- und Wunderbuch." This must date in part from the closing decades of the fifteenth century and in part from the early decades of the sixteenth. No articles have appeared in the literature on the Erlangen work; the Weimar text is discussed by E. Marx, "Bericht über ein Dokument mittelalterlicher Technik," *Beiträge zur Geschichte der Technik und Industrie, 16* (1926), 317-321.

[40] Prager and Scaglia, *Brunelleschi*, 128ff.

[41] Ladislao Reti, "Francesco di Giorgio Martini's Treatise on Engineering and its Plagiarists," *Technology and Culture, 4* (1963), 287-298.

[42] On the early phases of this development, see: Guilio Mandich, "Venetian Patents (1450-1550)," *Journal of the Patent Office Society, 30* (1948), 166-224; and "Venetian Origins of Inventor's Rights," *JPOS, 42* (1960), 378-382. Maximilian Frumkin, "The Origin of Patents," *JPOS, 27* (1945), 143-149; "Early History of Patents for Invention," *Transactions of the Newcomen Society, 26* (1947-49), 47-56; and "Les anciens brevets d'invention," *Archives internationales d'histoire des sciences,* N.S. 7 (1954), 315-323. Ramon A. Klitzke, "Historical Background of the English Patent Law," *JPOS, 41* (1959), 615-650. Hansjoerg Pohlmann, "The Inventor's Right in Early German Law," *JPOS, 43* (1961), 121-139. Frank D. Prager, "Examination of Inventions from the Middle Ages to 1836," *JPOS, 46* (1964), 268-291. Lyman R. Patterson, *Copyright in Historical Perspective* (Nashville, Tenn.: Vanderbilt University Press, 1968). Harry Ransom, *The First Copyright Statute* (Austin: University of Texas Press, 1956).

[43] Prager and Scaglia, *Brunelleschi*, 111ff.

for books necessarily involved a modicum of protection for their contents, although this was far from being an established part of sixteenth century law, custom, or attitudes.

In the midst of this checkered pattern of protection and non-protection, Besson's *Theatrum instrumentorum* appeared. It is prefaced by a lengthy royal privilege giving the author exclusive rights to the drawings and the ideas they embody and prohibiting their unauthorized copying or employment for a period of years. The utility of Besson's ideas to the common weal of the kingdom is given as the rationale for this step. I am unable to find in the available discussions of copyright history any mention of Besson's privilege, but it would seem to be one of the earliest of its type in France, particularly since it combines elements of copyright and patent. Ironically, it did not prevent an Englishman, Cyprian Lucar, from plagiarizing Besson for Elizabethan consumption across the Channel. Ramelli's *Le diverse et artificiose machine* appeared with only a brief, formal royal privilege in 1588; perhaps Ramelli thought his standing at court would discourage potential plagiarists. Some of his drawings were, in fact, expropriated by a member of his household, Ambroise Bachot, and published a year before Ramelli's book came out.[44] Ramelli complains in his preface about the theft and especially about the publication of his work under another's name and in "mutilated" form. He does not however invoke law or custom to punish Bachot, but merely expresses the hope that the discerning reader can distinguish between his own true drawings and Bachot's false copies. These episodes indicate that the relationship between authors, their material, and the audience was still somewhat unstable in the late sixteenth century, but they also show how far attitudes (of the authors at least) had evolved since the fifteenth century. Kings, legislatures and law courts would wrestle for many decades with the problems of creating an effective body of copyright and patent legislation; indeed recent events show that there are still unsolved problems in this realm. Nevertheless, the subsequent development of such legislation did incorporate many of the ideas and attitudes that first emerged among technical writers in the fifteenth and sixteenth centuries.

Magic, Science, Theology and Technology

One final cultural relation of technology needs consideration, its intellectual status or how it was seen to fit within schemes of knowledge. Again, there was considerable development in the two hundred years between Kyeser and Ramelli. The author of *Bellifortis* is outspoken in his views: the *artes mechanicae* include the *artes theurgice*, which rank just above the military arts in importance. Kyeser himself may have been a practicing sorcerer in addition to his astrological interests;[45] in any event, he organized his treatise around the major planetary signs, each of which governed certain arts. None of this is especially surprising. Technology had long been associated with magic in some of its forms, and the linkage would persist well into the seventeenth century. What is remarkable, in view of the utter conventionality of such views, is how little they were followed in either the German or the Italian schools. To be sure Kyeser's astrological symbols are reiterated in many of the works influenced by *Bellifortis*, but only occasionally do the signs preserve their organizing character. More often they are simply repeated out of context, and many texts dispense with magical overtones completely. Perhaps the drift towards military authors helped remove technology from its magical roots. The only persistent evidence of magical connections appears in the tendency of collectors of such manuscripts to bind them with other works on divination or natural magic. The Italians, at least after Fontana, show little inclination to assert a link between technology and magic; from time to time the "magical" properties of this plant and that animal will creep in, but there is nothing by way of systematic magic in the texts. Indeed, the early Italian authors seem quite uninterested in locating technology within any scheme of arts and sciences, except for Leonardo, who was always a bit more inclined toward theory than his predecessors. Leonardo's views on technology are not very clear, but his high respect for mathematics makes it seem at times that he considers mechanics to be related to geometry.

The connection at which Leonardo hints was to become a major theme of the sixteenth century works, particularly the "Theaters." For reasons that are not yet clear, the prefaces to such works run to great lengths to justify their appearance; thus they provide an excellent source for assessing attitudes toward

[44] Martha Teach Gnudi, "Agostino Ramelli and Ambroise Bachot," *Technology and Culture, 15* (1974) 614-625.
[45] White (see note 36), 438.

technology in the period. Besson's work, for example, though it contains no preface in its original printed version did have one in its manuscript form, and this has recently been studied by Alexander Keller.[46] Besson stresses the mathematical foundations of mechanics and seeks to provide the reader with general theories that underlie his inventions. The theory is somewhat disappointing to eyes accustomed to the glories of mathematical mechanics as expounded by Galileo and Newton; it is mainly conventional Aristotelianism with a bit of Archimedes thrown in. Still, though this may not be the font of the scientific revolution, it is interesting to note that the intention to link mathematics, physical theory and practical mechanics in one nexus is already part of the culture of technology by this date. Besson was a teacher of mathematics by profession, and one might be tempted to accuse him of expounding a purely personal philosophy were it not for the reiteration of many of the same themes by Ramelli and other writers. Ramelli's sense of mathematics is far cruder than Besson's (he reasserts, for example, medical astrology as a parallel case of the utility of applied mathematics) and his theory is far weaker, but he compensates by stressing at great length the public utility of new inventions.

Perhaps the most interesting aspect of the prefaces to the "Theaters" is the stress they almost always lay on theological justifications for technology. The mechanical arts are God's gift to mankind; they offer some compensation for the losses suffered at the time of the Fall. They alleviate, at least in part, the primordial curse of earning one's bread in the sweat of one's face. Properly practiced, they represent an imitation of the divine act of creation, and thus they are part of man's image-likeness to God as stated in *Genesis* 1. Just as God has ordered all things according to measure, weight, and number, so too the mechanical technologist arranges his creations according to the same principles.[47] Perhaps we would be unwise to make too much of these relatively conventional statements; many of them were derivative from scholastic thought and some may have been merely rhetorical "window dressing." Nevertheless, they seem to mark a definite attempt by technologists to win a recognized place in the order of the intellectual world by linking their activities to an undeniably important aspect of the high culture of the sixteenth century. They indicate that technology has moved away from its earlier associations with magic and theurgy and has become, at least in the minds of its articulate practitioners, an activity characterized by public utility, justified by the highest of theological considerations and rooted in the oldest of sciences, mathematics.[48] To this extent, the attitudes we usually associate with the seventeenth century were brought forth in the new body of technical writings from the preceding two centuries.

Conclusion

This outline of the relations between technology and culture is obviously a sketchy, preliminary attempt to indicate major areas where further work is needed. All the problems discussed above, the nature of technical writings, their authors and audiences, the philological problems they present, their connections with art, their legal history, and their relationship to science and theology--all these could benefit greatly from further consideration by specialized scholars. The history of early technology is largely virgin territory where important questions far outnumber significant answers. For example, one criticism that is frequently made of the works under discussion has been ignored in this paper, viz. that they often contain fantastic, impractical designs or devices. There is ample room for a study of technological fantasy as it began to emerge in these manuscripts and books; possibly such a study would tell us a great deal about the way western Europeans conceived of their arts and crafts in this period. Similarly, questions of cross-cultural comparison have been omitted. Is there any comparable literature outside Europe, and if so, how does it compare to the European style in these matters? The purpose of this paper has been two-fold: 1) to suggest alternative approaches to the history of early technology; and 2) to show that changes in the means by which technical data are stored and transmitted have implications for the

[46] Alexander Keller, "A Manuscript Version of Jacques Besson's Book of Machines, With His Unpublished Principles of Mechanis," in Hall and West (see note 30), 75-106.

[47] These and many other themes are analyzed in detail in Ansgar Stöcklein, *Leitbilder der Technik: Biblische Tradition und und technischer Fortschritt* (München: Heinz Moos, 1969).

[48] The exact nature of these transformations is yet another subject in need of further study. There seems to have been a continuous evolution of such attitudes from the medieval scholastic and moralistic writings to the documents under consideration. Cf. Stöcklein's monograph cited above and Peter Sternagel, Die artes mechanicae *im Mittelalter: Begriffs- und Bedeutungsgeschichte bis zum Ende des 13. Jahrhunderts* ("Münchner Historische Studien, Band 2;" Kallmünz: Lassleben, 1966).

interplay of influences between technology and the societies that foster technical achievement. We once thought that modern technology had its genesis in the period we liked to call the "Renaissance" and that it was somehow connected with the "revival of letters." We now know that the sub-literate craftsmen of medieval Europe gave birth to far more than the scribblers of later times; we also know that we were misled by the illusion to which Lynn White referred so that we mistook the documents for the activities they represent. Nevertheless, the scribblers had a contribution to make, for they help bridge the gap between the medieval technological accomplishment and its modern counterpart. If we are to understand the highly complex, but also highly significant process of technological evolution between the end of the middle ages and the beginning of the Industrial Revolution, then we must investigate in depth the cultural relationships surrounding technology in this period.

III. ART AND TECHNOLOGY

Proto-Astrolabes, Proto-Clocks and Proto-Calculators: The Point of Origin of High Mechanical Technology

by

Derek de Solla Price

There are tragically few really scientific objects of antiquity and we are looking for more. I suspect that the reason we don't have more is not because they are not there, but because we don't know what we are looking for. I would like to take an example of this where we can make the conjecture that there might really be objects which have not yet been identified.

We are going to deal with the origin of the astrolabe. The astrolabe, even if it is not a terribly important instrument itself—it is only a rotating star map, after all— is the one that spawned the craft. It is the instrument that was the chief product, chief training ground, and the masterpiece of work for a continuous tradition of craftspeople who preserved very special complicated techniques including scientific engraving. One would suppose that this fancy rotating disc star map would have started out in a simpler form as a sphere rather than the flat projection. So we all supposed years ago, before anyone knew, that spherical astrolabes would come before astrolabes. Then alas, eight or nine years ago, Francis Madison found the first surviving spherical astrolabe. It made us suddenly realize that it was the special sort of artifact that was alluded to in manuscripts, but that we just felt could not be.

The spherical astrolabe is a cap of open work with little pointers showing a star position and the whole cap can rotate on a grid of lines. As one turns the whole cap, it shows how the stars are sliding over the lines of altitude and azimuth. I suddenly realized when I saw this, that we had it back to front, and this was a late archaic adaptation like Gothic letters on modern tombstones, and that the real history began with a very familiar object, the star globe. We know that there were star globes with a cap sliding on the outside, and scales measuring the altitude and azimuth. It would then appear that the star globe is really the analog of the spherical astrolabe. The very oldest star globes have a little lobe on the top which is the typical throne of the astrolabe. The suspending device which one can see on the early astrolabes were very often three-lobed for reasons which will be described later.

It turns out that all of the earliest globes only look like globes when mounted on a stand. The legs are later additions, and all of the earliest specimens we have—early Medieval—are designed to be held in the hand by a little ring suspension on a throne. But even so, I don't think we are back to the beginning. What we really have to do is to explain the planispheric astrolabe. The astrolabe at the peak of its performance from the high Middle Ages is contemporary with the cathedrals. A fine example is the great Sloane astrolabe of the British Museum, a delight in workmanship, and one can see the network at the bottom and the open work rete above it, the form of the throne and the suspension. This may be traced back to the only Greek astrolabe, a Byzantine one, to be sure, now in Brescia. It is very much the same technique with throne, rete, and the network at the back with all the lines. At the back of the Byzantine astrolabe is a sighting device and shadow square, and in fact it seems that it is this back side of the astrolabe which became the mariners' astrolabes. They are famous because they are the very things that made the Spaniards such poor navigators. They provide observations that are three or four times worse than one can do without them, just using the naked eye. Insofar as they were used, no good captain would believe them. They were like the mariner's compass, and, indeed, all other instruments of navigation, in that they were inapplicable, and no good navigator really believed instruments until about the late eighteenth century.

This takes us back to the very earliest known astrolabe. It is early nineteenth century, and it is signed in Arabic "Bastulus" or "Vastulus" and clearly, this is the Arabic transcription of a Byzantine name that is very probably "Apostulus." We are very near the transmission into Islam and, in fact, the theorists tell us that this "Bastulus" is the apprentice of the person who began the craft. The first astrolabe makers of Islam were in the town of Haran. Haran—in upper Mesopotamia between Tigris and Euphrates, south of Turkey—is also the place where the Al-Jazar automation tradition arises, and where the French have excavated huge temples of the Sun and Moon, an astronomical cult very much like that of the island of Delos which was a sort of Switzerland, sundial factory of the ancient world. Haran is another one of these great centers of scientific astronomical worship. What I want to point out is that this is the earliest form of astrolabe we know, and yet if it goes back this far, it must go back to the Greeks. Unfortunately we do not have texts that take us back very much further.

We do know that we have planispheric projections. There is the famous Salzburg fragment, a fragment of monumental waterclock dial that is a star map, but not a star map of the cut-out variety, but a full-plate star map. It is a huge thing and must have been eight feet across its stereographic projection. We know that the Salzburg plate is not quite the first. A few years ago we discovered a little medallion of Antoninus—dating of about the second century A.D.—and we now have as many as three specimens. They are little, tiny sundials made in medallion forms as pocket pieces, and these dials are marked with an astrolabic stereographic projection, so the projection was then known. But where did they come from? Were there any astrolabes in deep Greek antiquity? A couple of years ago I was going through the Vienna Kunsthistorisches Museum and saw an exhibit of Greek sculpture and art work, and suddenly I said, "My God, a Greek astrolabe!" It looked very much like one, a container with a sort of hollowed-out space inside. It turned out to be in fact a Greek mirror, and the mirror had a little lobe on the side very much like the throne of an astrolabe, and it suddenly occurred to me that we should look at mirrors. I know what these looked like and I was aware of the corresponding very well known Egyptian tradition.

There is also linguistic evidence. It has always been a very well-known conjecture of astrolabe historians that calling the open work star map of the astrolabe *rete*, a spider's web, is wrong. It doesn't look like a spider's web at all. It is the tympan plate with all the lines and circles of stereoscopic projection on it that looks exactly like a spider's web. Could it be that the disc was engraved with something like that Salzburg disc with a set of constellations? On the handle of the mirror you would mount the spider's web, and turn the disc behind it by hand. Most significantly the little lobe that you hang the astrolabe up by is called in all languages "the throne." In the Arabic it is the *kursi*. Now is a throne, a seat, a chair something that you hang a thing from? Certainly not. Very clearly when the terminology was formed, the device was the other way up. And we have this particular form, even with the little two-lobed mount on the handle because the astrolabe was originally mounted on a handle like the armillary speres with a throne on which there sat the mirror of the universe—a disc marked out like the Salzburg one with pictures of constellations.

I have searched through all the published books I can find for pictures of Greek mirrors as well as Roman mirrors, knowing that some time or other I am going to come across a mirror with zodiac constellation decorations. In fact, it is not going to be a mirror at all; it is going to be the proto-astrolabe that I have now been looking for for about five years. Would you please help me find it? There are similar stories to tell but I have, after all, published them for the proto-calculators and for the proto-clocks like the Tower of Winds with all its embodied element theory.

I would much rather deal with any interest you have in questions, but I did want to do this one in particular to alert you art historians not only to look for Zodiac decoration in Greek mirrors, but also for any other objects that you spot either as scientific artifacts or an things technologically interesting. For, although all of man's early crafts and technologies are interesting, and each one is a life's work, there are occasional objects that enable us to jump in ten minutes, what would otherwise take a life's work of systematic effort.

Ancient and Medieval Gilding on Metal— The Technical Literature

by

S. M. Alexander

Several early texts in the field of art technology describe a number of procedures for gilding metal objects. Not all the practical methods of the craft are included, as a comparison between the written recipes and the surviving gilded materials readily shows. The written sources are nevertheless valuable for their descriptions of the procedures whose details sometimes increase our knowledge of ancient and medieval technology.

The Texts

Few ancient and medieval treatises are devoted entirely to metalworking. For the most part, written information on gilding is obtained from two types of sources:

1. Compilations of a general nature, such as Pliny's *Natural History*. Two of the thirty-seven books deal specifically with metals from several points of view, of which the technical procedures connected with metalworking are not the most important.

2. Technical treatises, existing from the first to third centuries A.D. onwards. They contain recipes, or more or less detailed descriptions of specific artistic processes, from which may be learned the actual materials employed, as well as the "scientific" and practical knowledge implied in their use. The most important of the technical treatises, with the information they contain on the gilding and other superficial coatings of metals, are discussed below.

a) The information in the earliest written recipes dealing specifically with the subject is doubtless of great antiquity, and probably originated in the activities of the priesthood and temple workers of Egypt and the ancient Near East. Its earliest significant and comprehensive written form, as far as may be judged today, was probably the 2nd century B.C. treatise of Bolos of Mendes, on the tinting of metals and stones, and dyeing in purple. This treatise no longer exists, but it is probable that the fragmentary *Physica et Mystica* of Pseudo-Democritus, the pertinent Papyri of Leyden, and the Stockholm Papyrus, all dating from between the 1st and 3rd centuries A.D., are partial copies of Bolos' famous work.[1]

The *Physica et Mystica* is a short treatise written in Greek, and is considered to be one of the earliest surviving sources in the history of alchemy. Its practical sections comprise individual recipes on purple dyeing and the manufacture of imitation gold and silver.[2] It may be a little earlier in date than the Greek

[1]R. Taton (ed.), *Histoire générale des Sciences,* Paris 1957, vol. 1, p. 379f.

[2]M. P. E. Berthelot, *Collection des anciens Alchimistes grecs,* Paris 1888, republished in one volume by Holland Press, London 1963. This contains the original Greek, a French translation, and commentary on Pseudo-Democritus and other early alchemical texts.

papyri from Thebes in Egypt, especially Papyrus X in Leyden and the Stockholm Papyrus, which together form an extensive treatise covering the entire range of topics recorded by Bolos of Mendes. For the purposes of this paper Papyrus X is the relevant work,[3] and specifically those recipes concerned with the surface treatment of metals whose composition is less than noble, to improve their appearance and to make them more nearly resemble gold and silver.[4]

A treatise entitled *On the most valuable and famed goldsmithing* exists in its most complete form only in a late 15th century manuscript.[5] However, an examination of its recipes shows a close relationship between several of the preparations and those of the Greco-Egyptian tradition. Berthelot is of the opinion that its formulae date from several periods, some of which may be as late as the 15th century, added at the time the manuscript was copied in its present form It contains 56 recipes devoted to the work of gold- and silversmiths including mercury gilding, parcel gilding, restoring faulty gilding, enamelling, engraving, with a few isolated recipes on writing in gold ink, polishing pearls, and the manufacture of soap.

b) Somewhat further removed in date from this early group of texts are a few treatises whose relationship with Hellenistic recipes is still evident, but which contain also additional formulae on many technical subjects, perhaps supplied by copyists of varying date. The *Compositiones Variae*, a unique manuscript usually dated to the 8th century A.D., is a source for data on the preparation of pigments and colored glass, as well as for a few aspects of metalworking.[6] The manuscript is written in Latin, but many of its recipes are evidently of Greek origin; some are derived from the same source as those of Papyrus X.[7]

Closely related in chronological and geographical origin are several recipes of the treatise entitled *Mappae Clavicula*, a more extensive and orderly work than the *Compositiones*.[8] Its numerous sections devoted to metalworking cover the same topics as those of Papyrus X in Leyden, with a few additional formulae from elsewhere. A single recipe for gilding copper occurs in the earlier section of the manuscript *De Coloribus et Artibus Romanorum* of Eraclius.[9] This work may be attributed to a 10th century author, perhaps of Italian origin, whose information is derived from earlier sources, rather than as a result of personal experience.

c) Eraclius Book 3 is one of the latest sources in which traces of the Hellenistic tradition are still clearly discernible. It is a varied collection of recipes from early Greek and Latin authors, with some original contemporary formulae, appended to the *De Coloribus et Artibus Romanorum*. Mrs. Merrifield considers it to have originated in Italy at a date shortly before the work of Theophilus,[10] that is, in the light of modern datings of that author, perhaps late 11th or early 12th century, rather than the 12th-13th century suggested by Mrs. Merrifield.

Theophilus' *De Diversis Artibus*, dating perhaps from the early 12th century, contains three books, each devoted to a different craft, namely painting, glassmaking, and metalworking.[11] Its ordered arrangement, precision of detail, and suggestions of personal experience make it the earliest of the "modern" handbooks on artistic procedure, as opposed to the compilations of the earlier copyists.

[3] Berthelot, op. cit., gives the history of the manuscript, a French translation, and an interpretation of its recipes. English translation: E. R. Caley, The Leyden Papyrus X, *Journal of Chemical Education* 3, 1926, 1149-1166.

[4] Papyrus X also deals with the manufacture of gold ink, the refining, testing and alloying of certain metals, and dyeing in purple. The Stockholm Papyrus has a few recipes on metalworking. Its main concern is dyeing and the manufacture of imitation precious stones.

[5] Berthelot, *op. cit., II, p. 307ff.*

[6] H. Hedfors, *Compositiones ad tingenda musiva,* Upsala 1932, with German translation.

[7] R. P. Johnson, "Compositiones Variae from Codex 490 in the Biblioteca Capitolare, Lucca," *University of Illinois Studies in Language and Literature* 23, no. 3, Urbana 1939.

[8] J. G. Hawthorne and C. S. Smith, Mappae Clavicula, A Little Key to the World of Medieval Techniques, *Transactions of the American Philosophical Society*, New Series vol. 64, Part 4, 1974.

[9] M. P. Merrifield, *Original Treatises in the Arts of Painting...,* London 1849, gives Latin text and English translation. Republished in Dover edition, New York 1967. J. C. Richards, "A new manuscript of Heraclius," *Speculum* 15, no. 3, 1940, 255-271.

[10] M. P. Merrifield, op. cit., vol. 1, introduction to Eraclius.

[11] C. R. Dodwell, *Theophilus De Diversis Artibus,* London 1961. J. G. Hawthorne and C. S. Smith, *On Divers Arts*, Chicago 1963.

The Procedures

1. Mechanical attachment of gold

a) Pliny[12] mentions the application of gold to a bronze statue of the youthful Alexander, by Lysippos, carried out on the orders of the emperor Nero (54-68 A.D.). As the artistic merit of the statue was not enhanced by this means, the gold was removed, leaving on the statue "scars from the work done on it and incisions in which the gold had been fastened." Pliny does not state whether the statue was wholly or partially gilded. His description suggests that the precious metal was used in the form of either thin sheet or gold foil, and attached to a statue of some size. As both "scars" and "incisions" are mentioned, it is possible that the gold was attached by means of rivets, which would leave a number of holes in several parts of the statue, depending on the size of the pieces of gold which were used for the gilding. Another possibility is that the surface of the statue was roughened or etched to provide a tooth for the foil, which was then attached by means of a suitable adhesive, or by burnishing the gold into the prepared surface of the bronze. The latter method would probably damage the statue more than the riveting. Whatever the precise procedure, it is to be noted that the method was selected at a time when mercury gilding was well known and frequently employed.[13]

b) A method of gilding silver thread is mentioned in the *Treatise on Goldsmithing*. The dating of these recipes has been noted as uncertain, but the following procedure seems to be of some antiquity. The silver, cast in a mould to form a long thin strip, is heated throughout its length two or three times. Its surface is then filed with a fine steel file, and gold leaf is rolled around the silver. The strip is heated to redness, removed from the heat, and rubbed with olive ashes. The process is repeated three times, then the strip is passed through the drawplate.

2. Lead-alloy method

Recipe 38 of the Papyrus of Leyden describes a method of gilding a copper ring so that it will resemble solid gold. Two parts of lead are ground with one part of gold to a powder as fine as flour. Gum is added as the medium, the copper surface is covered with the mixture, and heat is applied. The process is repeated until the desired colour (thickness) of gold is achieved. Berthelot considers that this was a very early method of gilding in use before a knowledge of mercury gilding was gained.[14]

3. Mercury gilding

Gilding by means of mercury is used only for objects of metal. According to the texts under consideration, the metals and alloys thus treated were electrum (a gold-silver alloy), silver, copper, brass, iron. The period of the introduction of mercury gilding is uncertain. The earliest literary references are found in Vitruvius[15] and Pliny,[16] but the earlier origin of the craft is attested by its description in Pseudo-Democritus, no. 4.[17] Pliny describes the process of gilding copper as follows: "The copper is first subjected to the violence of fire; then when it is red hot it is quenched with a mixture of brine, vinegar and alum, and afterwards put to a test, its brilliance of colour showing whether it has been sufficiently heated; then it is again dried in the fire, so that, after a thorough polishing with a mixture of pumice and alum, it is able to take the gold leaf laid on with quicksilver. (33.100) When ... things made of copper are gilded, a coat of quicksilver is applied underneath the gold leaf and keeps it in place with the greatest tenacity: but if the gold leaf is put on in one layer or is very thin, it reveals the

[12] 34.63. The text and translation used here are Pliny, *Natural History*, transl. H. Rackham, Loeb Classical Library (10 vols.) London 1952.

[13] P. A. Lins and W. A. Oddy, "The Origins of Mercury Gilding," *Journal of Archeological Science*, Vol. II, 1975. p. 365-373.

[14] Berthelot, *op. cit.*, I, 58. When the coated copper is heated the lead is converted to litharge, which then sublimes. The gold is held on the copper surface by a slight melting of the noble into the baser metal. Burnishing is possible to complete the operation.

[15] *De arch.*, VII.4

[16] *Nat. Hist.*, 33.65

[17] Berthelot, *op. cit.*, II, p. 46

quicksilver by its pale colour. Consequently persons intending this fraud adulterated the quicksilver used for this purpose with white of egg."

Pliny describes three stages in a five-stage process of mercury gilding:

1. the thorough cleaning of the copper surface by means, in this case, of strong heating followed by quenching in brine, vinegar and alum, and a polishing with pumice and alum
2. the application of mercury (quicksilver) to the cleaned copper surface
3. the application of leaves of gold over the mercury, in one or more layers.

He omits the two final steps:

4. heating to evaporate the mercury
5. the final burnishing of the gilded surface

The cleaning process of stage one is necessary in order to remove any grease or oxidation present on the surface of the metal that is to be gilded, and which would prevent the even penetration of the mercury, a process whose difficulty is not mentioned by Pliny. The cleaning described by Pliny would also etch the copper surface slightly. An even coating of mercury, on which the success of the operation depends, could be obtained by rubbing the mercury over the copper with a cork or other suitable implement; the gold leaf will not adhere to those parts of the object where the mercury is of insufficient penetration. Today the difficulty of this procedure is overcome by painting with a brush, on the metal to be gilded, a solution of mercuric oxide dissolved in nitric acid.

Pliny implies that the mercury is retained under the gold, revealing its presence by the pallid colour imparted to the gold as the two metals become amalgamated. The retention of the mercury would probably cause the gold to become powder and flake off.[18]

The method described by Pliny is repeated in Theophilus[19] and is mentioned by Biringuccio as being of current practice in the 16th century.[20]

A different method of mercury gilding, for a silver object, is contained in the Papyrus X of Leyden, no. 57. Mercury and leaves of gold are amalgamated to the consistency of wax. The surface of the silver is cleaned with alum, and is thereafter handled by means of a linen cloth to prevent soiling the surface with the fingers, and impairing the subsequent gilding. The waxy paste is laid on the silver in several layers. The mercury is then evaporated off by means of heat, and the object burnished. This method is repeated in the Treatise on Goldsmithing, the *Compositiones Variae*, the *Mappae Clavicula*, and Eraclius III.

A similar process, for applying silver to the surface of a copper vessel, is described at the beginning of recipe 25 in the Treatise on Goldsmithing.[21]

Equivalents of stopping-out varnish, where only partial gilding is required, are given in the Treatise on Goldsmithing and Eraclius III. These recipes are probably all of comparatively late date, but may reflect earlier practice. Surfaces which were not intended to be gilded could be covered with a mixture of

[18] cf. K. Kluge and K. Lehmann, *Die antiken Grossbronzen*, Berlin, vol. 1, ch. 5, part 3, p. 177 ff; also K. C. Bailey, *The Elder Pliny's Chapters on Chemical Subjects*, London 1932.

[19] Hawthorne and Smith, *op. cit.*, III.38

[20] *The Pirotechnia of Vannoccio Biringuccio*, translated by C. S. Smith and M. T. Gnudi 1942; MIT Paperback 1966.

[21] The method of attaching leaf over a prepared surface of mercury gives a more satisfactory result than the mercury-amalgam method. Less gold is required, it is easier to apply, and the burnishing is more readily accomplished. Modern gold leaf of double thickness, which is probably close to the thickness of the usual variety of Roman leaf, gives a more satisfactory result than the thinner leaf available today. In the process of burnishing (polishing), the joins between the leaves are evened out, so that it is not always possible to conjecture which process was used by looking for the edges of individual pieces of gold. It is sometimes possible to recognise the amalgamation method on small objects by the presence of two or three rough-edged layers of gold in places which are difficult to reach with a burnisher. In less than perfect work, very small ungilded areas may be visible, marking spots of insufficient penetration by the amalgam.

calcined bones, plaster, white lead and fish glue,[22] white of egg mixed with powdered brick,[23] lime and skin glue,[24] or soot (bistre) mixed with salt and white of egg dried over a charcoal fire.[25] When the gilding was complete, they were washed off the metal.

4. Surface enrichment

In the case of a "gold" object alloyed with silver or copper or base metals, to 14 carat or better,[26] it is possible to give a greater purity to the gold at the surface of the object by dissolving out the alloying materials.

The procedure is mentioned as early as the Papyrus of Leyden, whose recipe no. 20 gives directions for enriching the surface of a gold-copper alloy coin. The coin is alternately immersed in a solution of alum, salt and vinegar, and heated until the desired colour (i.e. purity) of the gold is attained. The principle was not necessarily intended only for fraudulent purposes, but was, and is, used to improve the colour of gold alloys.

Another recipe for the same purpose, Papyrus X no. 15, recommends that gold be immersed in a solution of *misy* (basic copper or iron sulphate).[27] salt and vinegar, with alternate heating and dipping. Recommendations for purifying silver in a similar manner are recorded in the *Compositiones Variae* through the use of burnt ox-dung, salt, or salt mixed with alum.

5. Substitutes for gold and silver

The application of various types of coatings to the less noble metals to make them pass for gold and silver seems to have had considerable currency in antique and medieval metalworking, judging from the frequency of the recipes devoted to these topics.[28] However, if they were used at all, many of the substitutes suggested in the recipes would have discoloured the metal in a short time, or would have worn away from the surface without leaving a trace of their presence. The two main groups are:

1. surface coatings which consist partly of a small amount of gold or silver, with a suitable coloured medium

2. materials which by chemical reaction or simple covering power turned the coated metal "gold" or "silver". Seen from a distance, some of them were probably moderately convincing. A selection of the recipes is given below:

1. The first category is represented by Leyden Papyrus X no. 28, where cumin, ground in water, is left to settle for three days. On the fourth day the mixture is shaken, and chrysocolla[29] is added to it, presumably in powdered form. The recipe ends here, but one presumes that the substance was painted on to the copper object to provide a yellow coating of poor adhesion. The coloured liquid from the cumin would give a more or less even yellow background against which the particles of metal could provide the necessary lustre.

A process of silvering, no. 81 in the Leyden Papyrus, calls for chrysocolla ground with white lead, earth of Chios,[30] and mercury. Honey is added, and the mixture is used for coating a vessel which had previously been cleaned with natron.[31]

2. The second category has a wider variety of suggestions:

22 *Treatise on Goldsmithing* no. 21.
23 *Ibid.*, no. 23.
24 *Ibid.*, no. 29.
25 Eraclius III.15.
26 H. Maryon, *Metalwork and Enamelling*, London 1959, p. 261.
27 See Hawthorne and Smith, *Mappae Clavicula*, recipe 1, note.
28 Gold-coloured alloys, such as copper-lead, were also recommended as substitutes for gold. For silver, copper whitened with arsenic is mentioned. Plating one metal with another cf. Pliny 34.160 and 162.
29 Gold solder, an alloy of gold, silver, copper. See Papyrus X nos. 31 and 33.
30 A kind of white clay: Berthelot *op. cit.*, I. p. 29, note 1. Cf. Pliny 35.56.
31 Cf. Hawthorne and Smith, *Mappae Clavicula* recipe 1, note 13.

a) "GILDING" SILVER

Recipe no. 51 in the Leyden Papyrus says that rubbing a silver object with a finely-ground mixture of *misy* (note 27), sandarach[32] and cinnabar[33] will give it the appearance of gold. No quantities are given.[34]. The *Treatise on Goldsmithing* no. 14 directs that equal parts of sulphur, garlic and lees be ground together, with the addition of dry lees, salt and urine. The preparation is heated over the fire and the object is placed in the hot solution for a set amount of time, followed by a quenching in cold water. The process is repeated five or six times.

An alternative, no. 35 in the same treatise, changes the ingredients to yellow curcuma[35] ground with dry lees and salt in water, otherwise repeating no. 14 above.

b) "GILDING" SILVER OR COPPER

Recipe no. 8 of Pseudo-Democritos states that cinnabar is made white by the addition of such substances as honey, vinegar, or alum, and then coloured yellow by *misy* or sulphur. It is applied to a silver surface to give the appearance of gold, or to copper to imitate electrum.

Leyden Papyrus no. 49 suggests the gilding of silver or copper by rubbing them with a solution of yellow natron[36] and salt in water.

Leyden Papyrus nos. 69 and 76 suggest *misy* and alum in equal parts, ground with one-third the amount of celandine, and made into a viscous solution with the urine of a child. The object is immersed in the solution until the desired colour is achieved. It is then heated and washed in cold water.[37]

On copper or silver figures or shields, a golden colour is imparted, according to the Leyden Papyrus no. 75, by a mixture of equal parts of orpiment,[38] copperas,[39] realgar, mercury, gum, and arum pith, made liquid with the bile of a goat.

c) "GILDING" COPPER

Pliny[40] states that thin leaves of copper used for theatrical crowns were painted with ox-gall to give the appearance of gilding. The method is repeated in Eraclius II.16, which source adds that the copper should be cleaned, then burnished with a bear's tooth before painting on the gall. The gall is applied in three coats.

Recipes for "gilding" copper are quite numerous in the early medieval texts, where such substances as realgar, orpiment, gum tragacanth, goat bile and egg yolk are applied over a metal surface previously coated with oil. Similar processes are found in the Treatise on Goldsmithing, eg. no. 55.

d) "GILDING" TIN LEAF

Tin is frequently suggested as a substitute for silver in various crafts. It is probable that gilded tin leaf was intended for use in manuscript painting rather than in metalworking, although it may also have found some use in the latter craft.

32 In this instance probably realgar, As_2S_2.

33 Mercuric sulphide HgS.

34 The recipe is for a surface coating which leaves a faint gold colour on the silver when the excess is removed.

35 Turmeric, a dye plant, "Indian saffron."

36 Cf. Pliny 31.46. Both metals assume a golden translucent colour for a while, then turn black. The coating seems to adhere better to silver than to copper. Perhaps these recipes suggest a preliminary cleaning of the surface before the actual gilding takes place.

37 Copper immersed for 14 hours beforehand in the urine takes the colour better than either the copper or the silver surface manipulated as in the recipe. The resulting colour is dull yellow.

38 As_2S_3.

39 Green vitriol, protosulphate of iron.

40 34.94.

The leaf was coloured with saffron and glue dissolved in water, and painted on the heated metal; or with saffron, orpiment and gum, ground and boiled with linsed oil and rain water; or the leaves of tin were varnished with three or four coats, then immersed in a boiling solution of beer and bistre (Eraclius III.13). This last process is recommended also for restoring areas of damaged gilding on metal vases.

e) SILVER COLOUR FOR COPPER

The simplest method for giving the appearance of silver to copper is described in the Leyden Papyrus no. 42. The cleaned copper surface is rubbed over with mercury.[41] Alternatively, according to the same recipe, the cleaned copper object could be immersed in mercury mixed with white lead.

Another method was to form an amalgam of tin with mercury, as in recipe 27 of the Papyrus, for applying to the copper surface.

A related recipe, nos. 67 and 80 in the Papyrus, describes a mixture of cinnabar with alum and Cimolean earth[42] and sea water, to give the colour of electrum, presumably to a copper object.

* * *

Certain features are at once apparent from the above survey of gilding recipes:

1. Gilding techniques of surviving antique objects have nothing in common with the techniques described in the late antique "technical" literature as represented by the *Physica et Mystica*, Papyrus X, and certain recipes in the *Compositiones Variae* and the *Mappae Clavicula*.

2. A few of the gilding techniques of surviving antique objects are described in the "general" as opposed to the "technical" literature. The "general" literature is classical writings as a whole with their chance references to metalworking and other techniques in the arts.

3. Most of the gilding techniques of surviving antique objects are not described in any early literature. We know from such authors as Pliny and Vitruvius that several Greek painters and sculptors wrote about their work[43] but no mention is made of any author writing on metalworking. It seems probable, however, that such treatises were written, because although the earliest texts mentioned in this paper are clearly alchemical rather than artistic in their content, the later compilations such as *Compositiones* and the *Mappae Clavicula* have some recipes which are applicable to the arts. Presumably these were copied from texts other than those devoted to alchemy.

[41]The surface can be polished with cotton wool. It retains its lustrous appearance for a few hours. When the mercury begins to penetrate further into the copper, the lustre is lost.

[42]Cf. Pliny 35.195ff.

[43]Towards a History of Art Materials, *Art and Archaeology Technical Abstracts* vol. 7, no. 3 (supplement). There is much to suggest that these writings were other than technical handbooks.

Persian Locks as Works of Art

by

John Alden Williams

Anyone who examines the Parviz Tanavoli collection of locks from Iran will be struck by the extent to which the lockmakers have employed skill, taste, and creative imagination in what they have produced. Some of the locks were worn as charms, and others were suspended like ornaments on boxes they secured, so it is understandable that these are in themselves ornaments as well as objects of everyday use. What is less apparent is the way that most of them have been treated as *objects to be ornamented*, and thus to be enjoyed as well as used.

It is therefore a worthwhile exercise to consider these useful artifacts as objects of art, and to attempt to classify the ornament that we find here, to see if general patterns become apparent, and to look for analogies elsewhere in the art of the Islamic peoples.

There is only one lock here that is bare of any attempt at decoration on its surface, but it has a beautiful, sculpted shape: the silver lock identified as coming from an Isfahan shrine, and dating from the 15th or 16th century. The locksmith has also been a successful silversmith; like a sculpture, its sensuous, pleasing curves draw our attention to the qualities of its material. This lock is in a class by itself. However agreeably it may impress us, it is quite atypical of the other locks, or of the general tendency of Islamic objects of art. It exalts the material from which it is made, for one thing; it has a quality that is reminiscent of fine European table silver (plate XXIII, ill. 1). Its shape is so unusual in Islamic metalwork, is to be most easily explained by the contacts that Western Iran had with the Venetians and with other Europeans at this period, particularly in the period of the Whitesheep Turkoman State in the late 15th century. It is undeniably artistic, but there is little in the artistic tradition of Iran to resemble it. It is much more typical of the Persian locksmiths who made these objects to make them look like something else: a handbag, a lion, a man, a bird; to cover them with decoration so as to mask the surface. The artisan does not say, "See, this is silver," but, "Look again. Things are seldom what they seem." It may support my hypothesis about this lock to point out that it has a locking mechanism which is unlike that of any other lock in this collection. I surmise that it was either made by a European, or by someone (e.g., an Armenian Christian) trained in a European atelier.

Much more typical are the locks (in silver as well, usually) which date usually from the 19th century and repeat a familiar motif in Persian textiles, the *boteh* or Kashmir pattern, like a comma. The form is so familiar to us that there is no need to pursue it, but the form is one particularly found in the Persian tradition of Islamic art, which spread from Iran to Kashmir and then to Paisley, in Scotland.

Also in the category of pleasing and not necessarily meaningful shapes are some brass locks, apparently from Isfahan, dated between 1600 and 1700, that is to say from the period dominated by the art of the court of Shah 'Abbas the Great. Similarly rich curvilinear designs are especially found in the work of the Timuri renaissance of the 15th century, from which the Safavis derived them. As in the case of many of the 19th century *boteh*-forms, however, these Shah 'Abbasi forms are usually chased and decorated on the surface, unlike the silver shrine lock (plate XXIV, ill. 2).

There are also talisman locks of the 18th and 19th centuries, sometimes inscribed with letters and words, which have attractive shapes. They were designed to be worn as charms, for example by pregnant women, so as to magically appropriate the locks' properties of binding and loosing for the wearer, and it is easy to see why the makers would wish to give them gracious shapes, so that they might also be an ornament (plate XXIV, ill. 3).

A treasury lock, dated 1500-1700, i.e., from the Safavi period in Iran, has a fine, business-like shape (plate XXV, ill. 4). But it is inlaid with stars and palms. Palms figure often in the Near East as symbols of victory: for example palm-branches are often carried in processions commemorating Muslim saints and martyrs, yet they are not often depicted in Islamic decoration. Nothing, however, is more natural than to see stars in Islamic art: they are ubiquitous. The star has a special significance for Muslims: quite early in their history they began decorating both places of worship and objects of everyday use with stars. They recall many passages in the Qur'an: e.g., "He it is who appointed for you the stars, that you might be guided by them in the darkness on land and sea." (VI, 97) "God has cast upon the earth firm mountains, lest the earth shake with you; and rivers and ways, so that haply you may be guided. He has appointed them as waymarks for you, and by the stars are men also guided. Is He who creates like unto one who does not create? Will you not remember?" (XVI, 16) In short, stars remind man that the signs of God's guidance are all around him, in the real world. They are familiar reminders that God guides mankind by His prophets, and by Scripture, and indeed by everything around us, if man will but see. And if we ask "Very well, but why put them on a treasury lock?" the answer might be "So you think that people who handle money don't need guidance?" It is certainly as good a place for them as anywhere else. Given the frequency of stars and starpatterns in Islamic decoration, it is only surprising that relatively few of the locks in this collection bear them.

There is a whole class of locks which are actually little animal sculptures. Many of them date from the Seljuk period, i.e., 1000 to 1200 A.D., such as lions and horses. The Seljuks were Turks, and had used animals as totems in Central Asia before they became the ruling class in Persia. Some people have tried to trace the interest in animal art to the Turks; others have pointed out that the native Iranians themselves had old traditions of animal art, and that the Zoroastrian scriptures command men to be sympathetic with animals, because they are also the children of "Our Mother, the Earth." It is certainly true that there was a lively interest in animal art in Persia, but what is even more significant, perhaps, is that it was taken up by Islamic civilization. A lock made like a humpback zebu cow, or perhaps a buffalo, dating from the Seljuk period, is paralleled by ceramic jugs with cow-head spouts from pre-Seljuk Islamic Iran, in the 9th or 10th century, and 10th century lusterware from Baghdad with painted zebus. Is it necessary to mention here that old canard: "The Art of Islam forbids the representation of living things?" One needs only to look. How it has been possible for people to so deny their eyes as to go on repeating this obviously false information for years—so that until now one finds it in general art history books—one can scarcely guess. One can even hear Muslims, who ought to know better, say that the Qur'an forbids the representation of men and animals; yet it does nothing of the sort. Only some dubious statements attributed on shaky authority to the Prophet Muhammad are critical of artists. No one took them seriously, for the artists of Islam went right on through history depicting living things, men and animals. Here is a lock carved like a horse from Seljuk Iran, and there is a rabbit (plate XXVI, ill. 5), carved in 11th century Egypt—where there were, by the way, neither Turks nor Persians in significant numbers—apparently simply as an *objet de vertu*. A lock carved like a lion, from Seljuk Persia, is similar to a bronze lion with a serpent tail from 11th century Egypt (plate XXVI, ill. 6)—use unknown: it was probably a small aquamanile, or a water pitcher, and is very similar to some of the locks in this exhibition, in its shape. In their way, all the Muslim peoples seem to have been sympathetic to animals—in fact, a group of traditions from their Prophet, on good authority, recommends them to be. Certainly they had no hesitation to display their sympathy in depicting them.

Several of the locks from the Seljuk period, and later as well, are shaped like rams; a favorite animal in pre-Islamic Persia, and a symbol of royal authority. I do not know of any other reasons why they might have been favored by lockmakers, over the obviously protective lion or bull. We can compare them to the sculpture of an antelope, from 11th century Cairo: rather incongruously, it is covered with coiling vegetable ornament (plate XXVI, ill. 7). Islamic artists rarely chose to leave surfaces unornamented, in the high periods of Islamic arts, and we may note that many of these animal locks also have arabesques of vines, or bits of geometric ornament on them, as well. A void is to avoid: the empty place is the place where God's creative activity is not being manifested. For Muslim aesthetics, emptiness is ugly, and less is seldom more.

A particularly handsome silver lock from the 18th century, probably, or the early 19th, is engraved with two birds on either side of a plant, which bears a flower. This is a variation of a very, very old motif in Near Eastern art; the tree of life, guarded by protective genii (plate XXVII, ill. 8). The theme is ancient; in Sassanian textiles it is often the king who figures on either side of the tree, a guardian of the vital principle. But the decorator of the lock has rung a change on the old theme; we have two nightingales affronting a rose. The nightingales are popular symbols of the lover—that is, Man the mystic; the rose, recipient of the nightingale's hopeless lovesong, is the eternal Beloved: that is God—and it is likely that the scene is intended to remind us that the adoration of the eternal beloved brings life. Such sentiments became the clichés of Persian Islamic culture; they may seem to us exalted, but to the sort of people who used these locks, they were everyday truths.

There is a series of steel locks from Shiraz, the City of Nightingales, which have the shape of birds, and reputedly these are the beloved bird of the Shiraz poets. As a symbol of the lover, they obviously are well suited for attachment or suspension to treasured objects, e.g., on boxes or cupboards for keepsakes or jewellery.

Lions were always popular for locks. Lions were found in Persia until quite recently; there may even be a few left in remote areas, and they were regarded, properly, as both magnificent and forbidding: just the thing to guard one's valuables, whether in the 11th century or the 17th. We may note, as an aside, that there are probably more sculptured representations of lions in Islamic art than there are of people: statues of lions seem often to have been placed in front of gates and palaces.

With some of the locks of the Tanavoli collection, dating from the Zajar period—i.e., from the 18th and 19th centuries—the face of a lion is shown, given cut-glass eyes and a beauty spot or *khal*, on the forehead (plate XXVII, ill. 9). The mane of the lion radiates,
we might say, around the face, and reminds us of the way that the sun is often depicted, with a face, e.g., on the 17th century Shir-Dar Madrasa of Samarqand (plate XXVII, ill. 10). What we seem to have—though to be sure, not in any usual iconography—is the *Sun in Leo,* the zodiacal sign of kings, which figures on the national flag of Persia; a potent reminder of the importance of the monarchic tradition there from ancient times.

Doves and fish also appear among the figural locks, made in steel, brass or bronze and silver. One of the earliest locks in the collection is a pre-Islamic bronze fish. What it meant for the pre-Islamic Persians I cannot say, but most of the fish locks occur from the 17th and 18th centuries, and by this time the fish was well established as the sign of Pisces, which is always shown accompanying Jupiter, the Ruler, in the astrological iconography appearing on ceramics, metal, portals, and elsewhere. Its association with governing, domination, might well recommend it for a lock form. Doves, of course, were associated with various goddesses of the ancient Near East, and have to do with affections of the heart.

A very interesting group of brass figural locks, sand-cast, flat and thin, with curious shapes, has been ascribed by the authors of the catalogue of the Tanavoli collection to a period between the mid-11th and mlid-13th centuries.[1] The known examples of this group include a centaur shooting a bow, the sign of Sagittarius; a creature with the scales and body of a fish and the forepaws, torso and head of a quadruped, the sign of Cetus; a human-headed bird, or harpy; a lion; a lion with human face and beard; an eagle; a man riding a horse whose hindquarters are a fish tail; and a male figure in coat and knee-breeches (plate XXVIII, ill. 11). It is on iconographic grounds that the ascription to the Seljuk period has

[1]Tanavoli, Parviz and Wertime, John. 1976. *Locks from Iran.* Washington: Smithsonian Institution, pp. 88-90.

been made, because harpies and representations of Sagittarius and Cetus were all fairly prevalent then, and these locks clearly belong to the same series. These arguments however are not persuasive. A Sagittarius appears on the gate to the 17th century Qaysariya of Isfahan, and harpies, sphinxes and other monsters appear on Iznik pottery of the same century. Nothing is more common in Islamic art than the transfer of decorative motifs from one medium to another, or from a small object to a large monument; it is one of the peculiarities of the tradition. Legendary creatures of this sort are by no means confined to the centuries where the authors wish to assign the locks. The human figure in knee-breeches and coat is however much more distinctive of a fabulous monster well known in Persia in the 17th and 18th centuries: the European. It is most likely that this series of engraved brass locks was made in the 17th century, and probably at Isfahan.

Some of the locks are shaped like human heads. Human heads are used in early Islamic art, e.g., in the bath at Khirbat al-Mafjar, ca. 740 A.D., and were probably taken from Iran, where in Parthian art disembodied heads may have first signified gods. By Islamic times, they seem to have been merely a convention, though to be sure one which must have been continued because of feeling in Muslims. In the Anatolian tomb of a Muslim princess from Nigde, dated 1321, we can again see a head used as an ornament (fig 12). It is difficult to say what this meant to the ordinary Muslim beholder—in Persian manuscript illustration, where hidden heads peer from trees and rocks, it seems to be a reminder that the world is alive, created and sentient, and at every level of nature reflects its Creator. In any case, again we can see that Islamic art is not humanistic; the human form is not its predilection, but it reproduces the human face or form often enough as a decorative element.

We pass now to those locks which are ornamented, rather than being treated as ornaments in their own right. A group of bronze rectangular locks from Western Iran in the 19th century is treated with interlocking and concentric circles. This is a very old motif in Islamic decoration, so rich in all manner of all geometric motifs; it occurs for example on mosaic floors in palaces in the Syrian desert in the first century of Islam, and comes, like the mosaic technique itself, from the Late Antique. However, the locks of the Tanavoli collection are not distinguished for their geometric ornament: it is rather as if the locksmiths have paid a perfunctory service to a prevailing taste: the geometric motifs that appear on the locks are not done with the same enthusiasm as the sculpting of animals. Still they appear regularly enough to be noteworthy. Interlaces appear on some of the earlier locks, and other repeating patterns. We have already called attention to the relative scarcity on the locks of starpatterns, which are so often used in Islamic geometric decoration. The mathematical harmony and ordered regularity of geometric motifs reflect a central conviction of Islamic civilization, that all creation is a harmonious and perfectly arranged work of marvelous complexity, well ordered by a Supreme Designer.

Another class of locks was decorated with vines and floral motifs, usually continuous. The Muslims had lifted the vine motif from the art of early Christianity, where it recalled the saying of Jesus, "I am the Vine; you are the branches," i.e., expressed the unity of Christ with His Church, and reminded the initiate of the sacrifice of the Cross, unbloodily repeated at the altar. To the Muslims, the vine was simply the evidence of God's abundant provision for His creatures; to remind them that He is the King of the Universe, who causes the barren earth to bring forth the vine. Along with the cornucopia, it appears on some of the earliest Islamic monuments, and in the *mihrabs*, or prayer niches. Unlike early Christian art, Islamic art was very chary of symbols, though full of significance. Because Islam refused to be committed to religious symbols, it developed ornament which was almost secular, and the same techniques that were used for the most sacred object could also be used for the most profane. A vine, after all, is only a vine, if you will see nothing more.

One lock, decorated with floral motifs, has on its back the words *Mawla Husayn;* an ambiguous phrase, used by Shi'i devotees, which could mean either "Lord Husayn" or "This is Husayn's client." Perhaps this was one of the locks once worn in a cut in the flesh of one of the fasting penitents on Ashura or Tenth of Muharram, the Shi'i Day of Atonement.

This brings us to writing as ornament. A number of the locks are inscribed with letters and/or numbers; let us exclude these from our discussion, for it is clear that we are not dealing with ornament here at all.

These are talismans, and the lettering is supposed to be a magic spell; that is, it is functional, not ornamental, in intent. What ornament there is is to be found in the shape of the lock itself. All of these inscribed talismans are quite late, and most of the talisman-artifacts I have seen from the Islamic world seem to be late; they tended to proliferate in the 18th and 19th centuries when Islamic civilization was no longer flourishing, when Muslims felt threatened by forces they could not control, and even doubted at times that God was completely in charge of affairs. One might go so far as to say that such talismans are not usually products of Islamic civilization at all; they are articles of post-Islamic superstition from the period of universal decadence that fell on the Islamic world in the 18th and 19th centuries.

Other locks however are inscribed with writing clearly intended to be ornament. Because most of the inscribed locks are Safavi or post-Safavi, they favor the script most prominent in Persia in that period, the cursive script called *nastaliq*. But at all times in Islam, writing is seen as a beautiful thing, and the calligrapher is the most honored of all artists. If Christianity developed a humanistic art, it was because the divine Word was made flesh. But in Islam, the divine Word was made writing. Though 85 to 90 percent of the population might be illiterate, they found that writing was significant and appealing; at times even the appearance of writing was used as ornament. False scripts appear in folk arts such as tribal carpets, and there is even something that looks like script but is not on a fine shrine lock from the Qajar period, alongside genuine inscriptions (plate XXVIII, ill. 13). God's most perfect revelation of Himself was Scripture: all else, stars, arabesque, geometric, animal or human art, is only an aspect of that truth. At times calligraphy may be combined with these, at times it stands alone.

Hence these artifacts, when considered as objects of art, demonstrate general principles that seem to apply to all Islamic art. The decorative vocabularly is not specific; it applies as well to a carpet, a box, a tray, or a mosque. It avoids the unique, for the most part; the artist's role is not that of a prophet, but of a virtuoso performer, whose role is to please, and to make life attractive and enjoyable while reflecting revealed truth. Rather like the many poems of this culture which celebrate a love that may be equally sacred or profane, this art is all ultimately religious, with an extensive secular vocabulary. It seems altogether appropriate for a civilization one of whose great saints had this to say of saintliness in the 11th century a.D.: "The true saint sits in the midst of his fellowmen, and rises up and eats and buys and sells and gives and takes in the bazaar among other people, and marries and has social intercourse, and never for an instant forgets God." (Abu Sa'id ibn Abi al-Khayr of Khurasan)[2]

[2] Smith, Margaret. 1950. *Readings from the Mystics of Islam*. London, p. 49.

PLATES

Ill. 1. Sundial pot.

Ill. 2. 15th century version of the ancient armillary sphere.

Ill. 3. Sundial on the Acropolis of Athens.

Ill. 4. Earliest known Astrolabe, made by Bastulus.

Ill. 5. Monumental Stardisk.

Ill. 6. The Tower of Winds in the Roman Agora in Athens.

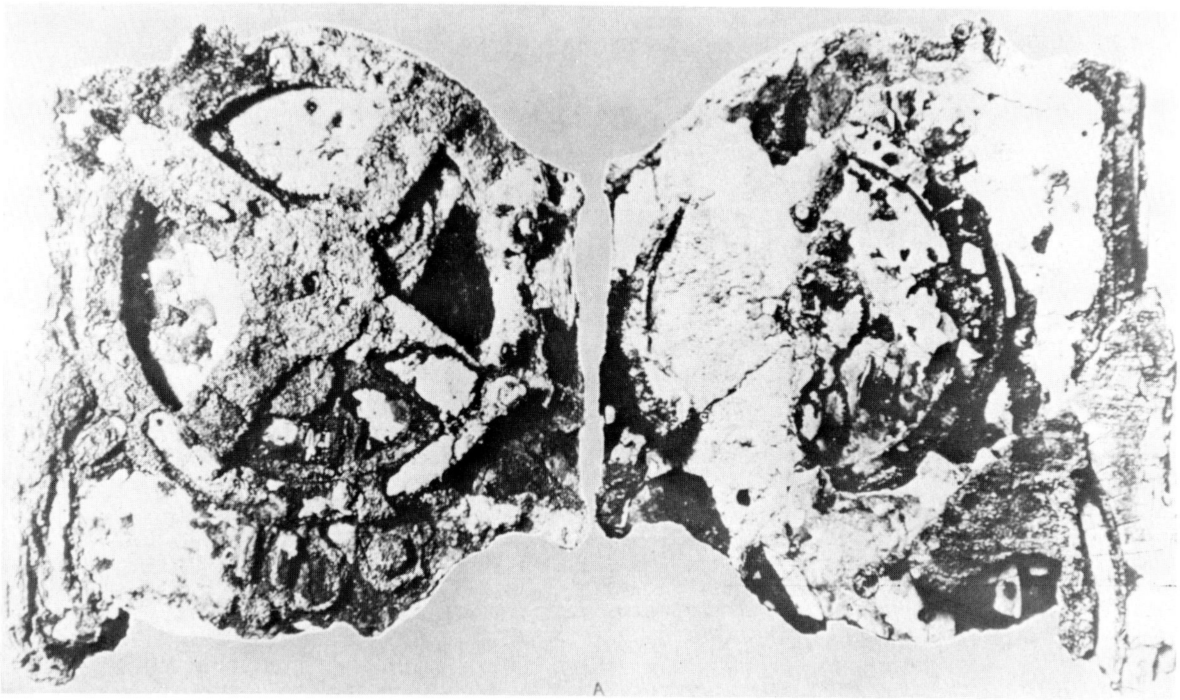

Ill. 7 a, b. Antikythera Mechanism.

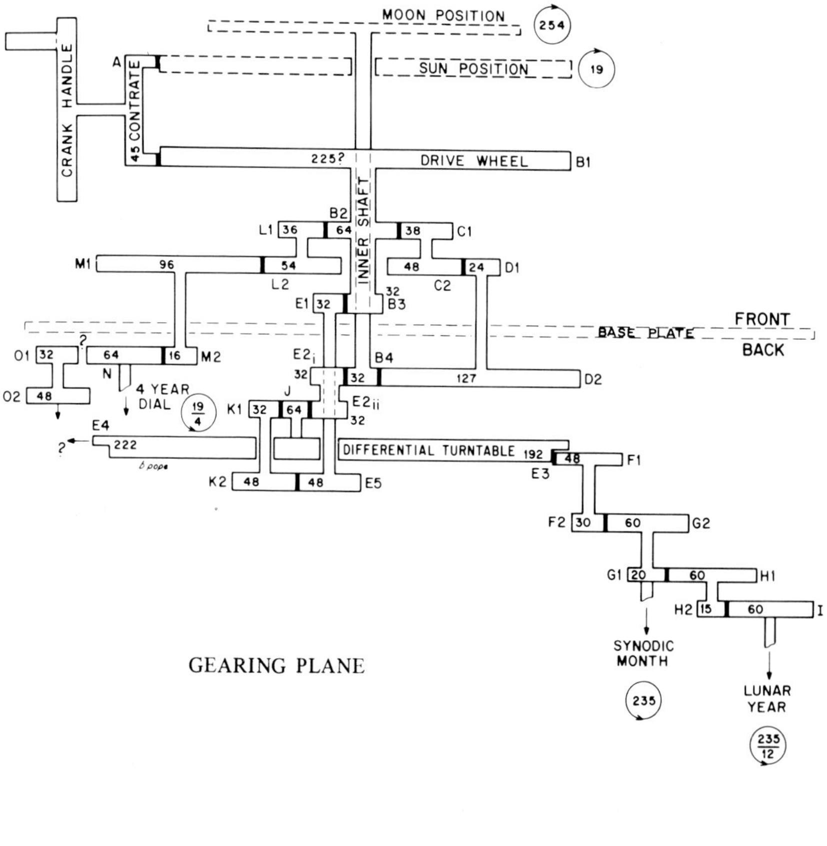

GEARING PLANE

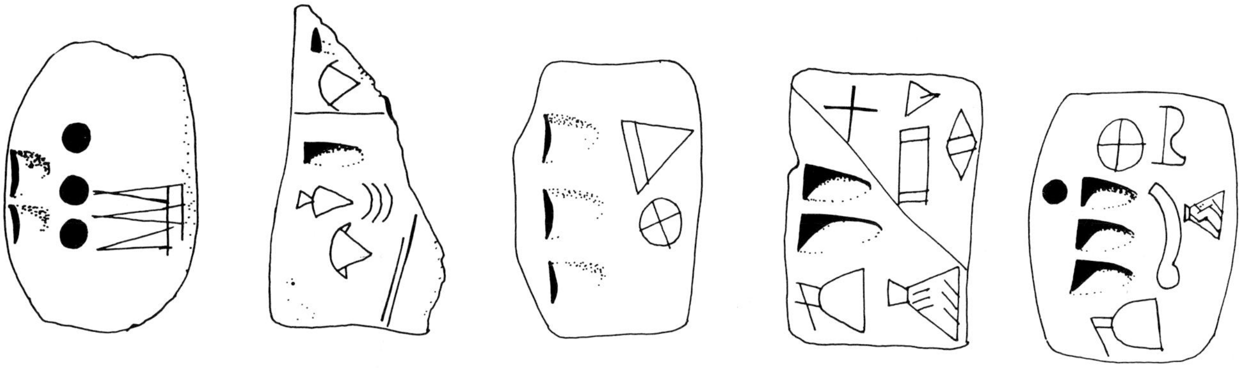

Ill. 1. Clay tablets from Warka.

Ill. 2. Land sale.

Ill. 3. Clay tablet with impressed signs from Susa.

Ill. 4. Clay envelope (*Bulla*) with tokens contained inside from Susa.

Ill. 5. Tokens from Susa (3300 B.C.).

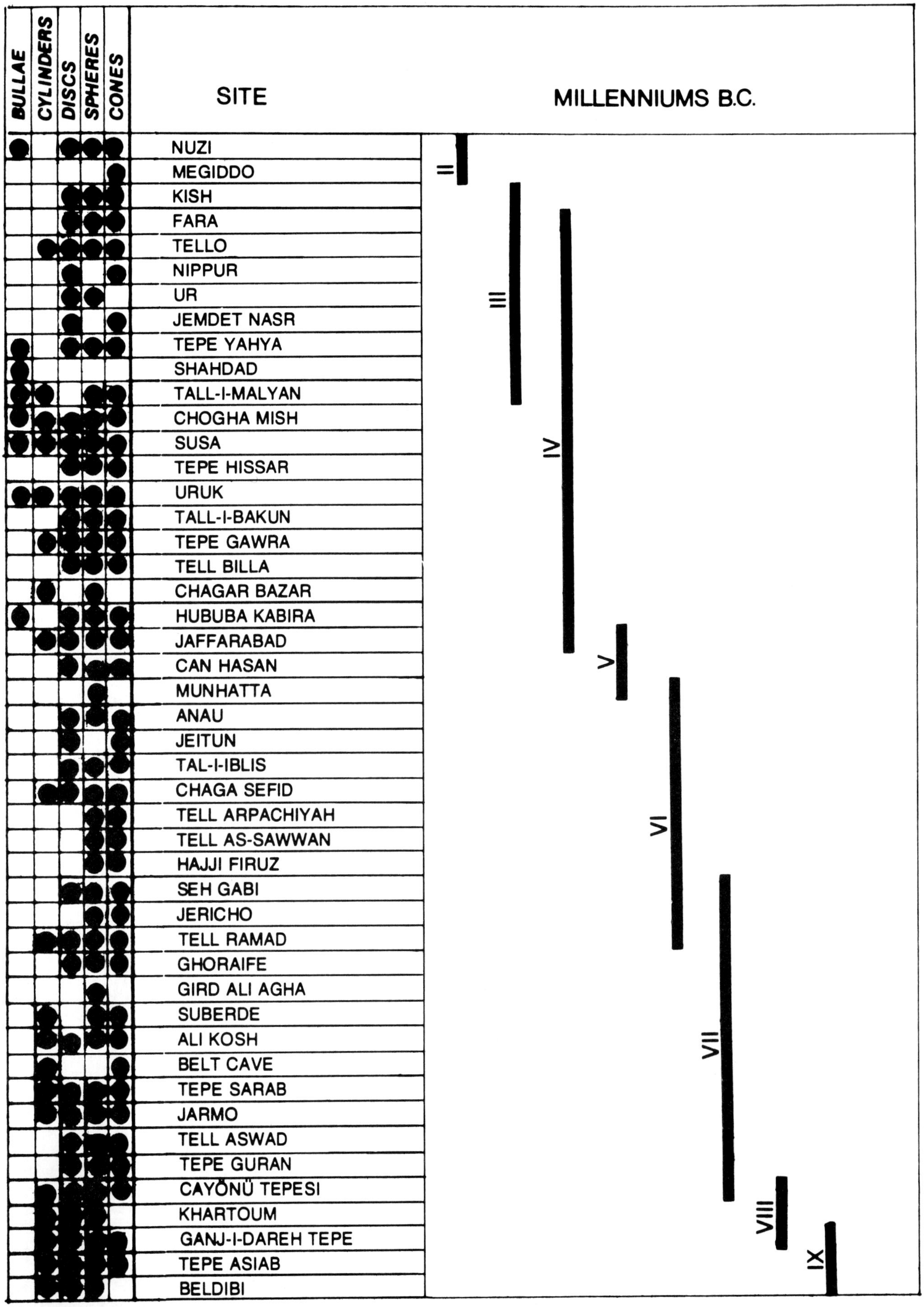
BULLAE
CYLINDERS
DISCS
SPHERES
CONES
SITE
MILLENNIUMS B.C.
NUZI
MEGIDDO
KISH
FARA
TELLO
NIPPUR
UR
JEMDET NASR
TEPE YAHYA
SHAHDAD
TALL-I-MALYAN
CHOGHA MISH
SUSA
TEPE HISSAR
URUK
TALL-I-BAKUN
TEPE GAWRA
TELL BILLA
CHAGAR BAZAR
HUBUBA KABIRA
JAFFARABAD
CAN HASAN
MUNHATTA
ANAU
JEITUN
TAL-I-IBLIS
CHAGA SEFID
TELL ARPACHIYAH
TELL AS-SAWWAN
HAJJI FIRUZ
SEH GABI
JERICHO
TELL RAMAD
GHORAIFE
GIRD ALI AGHA
SUBERDE
ALI KOSH
BELT CAVE
TEPE SARAB
JARMO
TELL ASWAD
TEPE GURAN
CAYÖNÜ TEPESI
KHARTOUM
GANJ-I-DAREH TEPE
TEPE ASIAB
BELDIBI
II
III
IV
V
VI
VII
VIII
IX

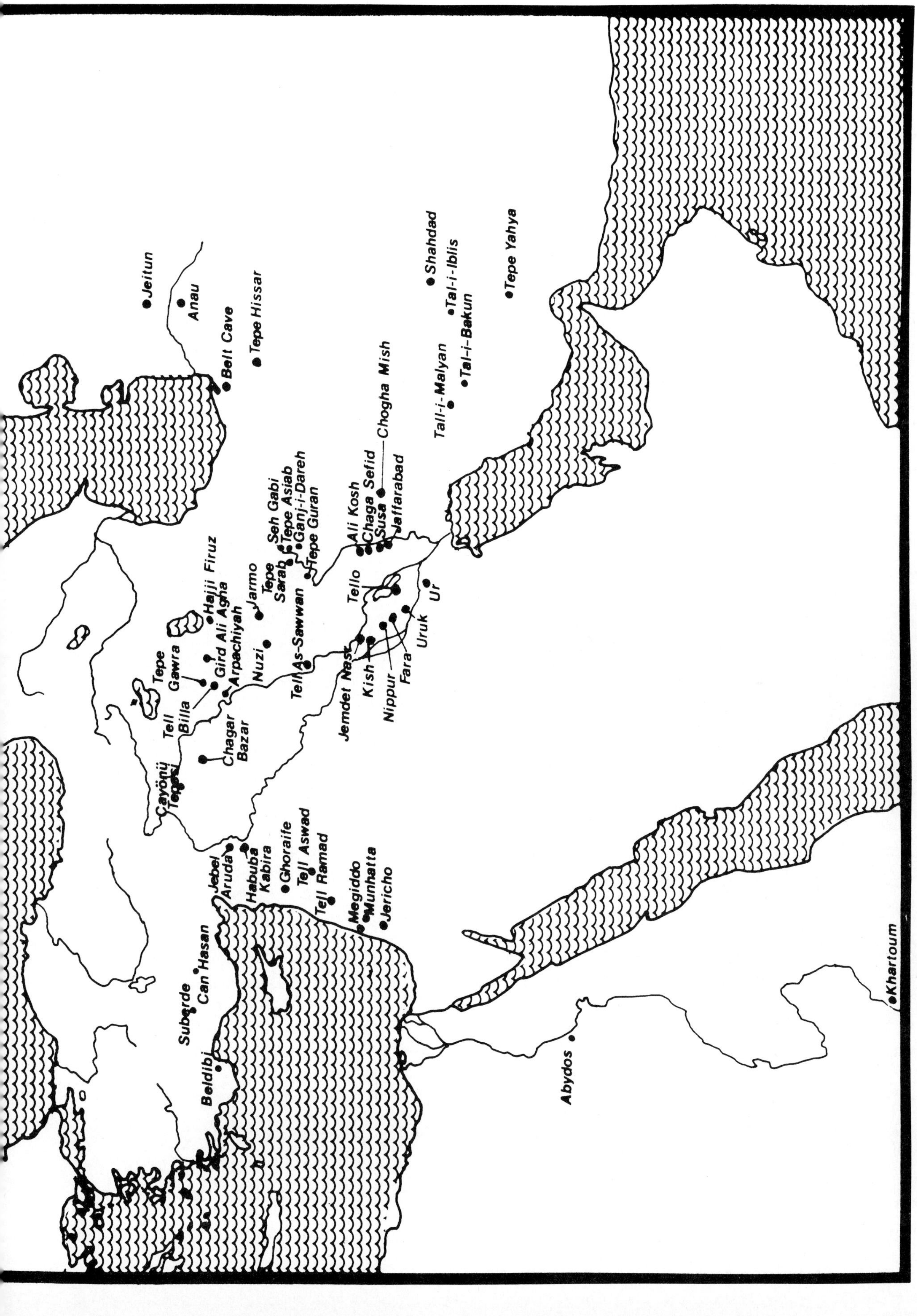

Ill. 7. Geographical distribution of tokens.

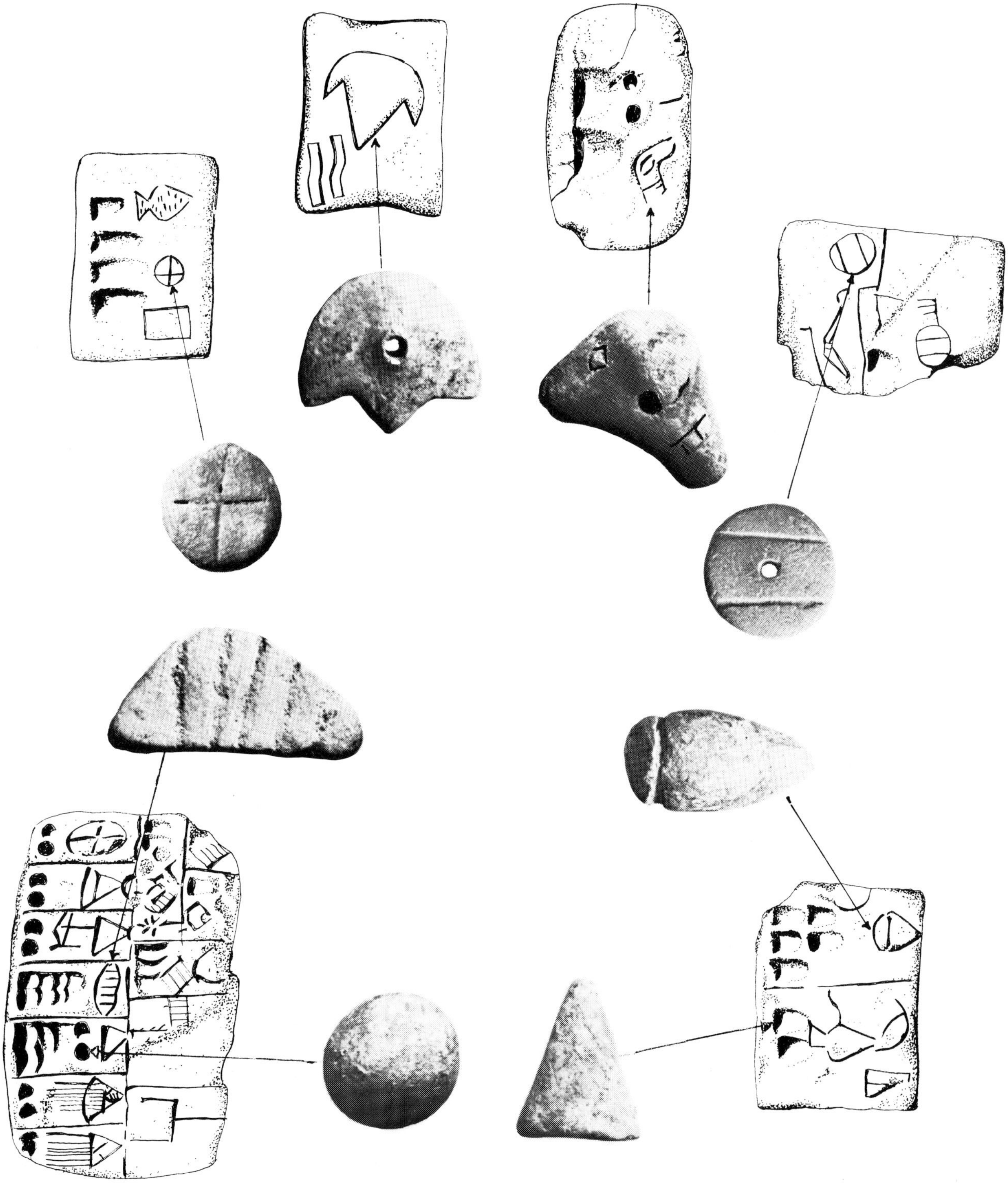

Ill. 8. Illustration of similarity between tokens and pictographs.

Ill. 9. Token assemblage from Susa.

Ill. 10. Clay envelope (*Bulla*) with impressed signs corresponding to the token content from Susa.

Ill. 1. Pictorial dedication of a textless manuscript to Emperor Sigismund, ca. 1435.
In the original, the scene here depicted is repeated on the open page of the book being handed to Sigismund.

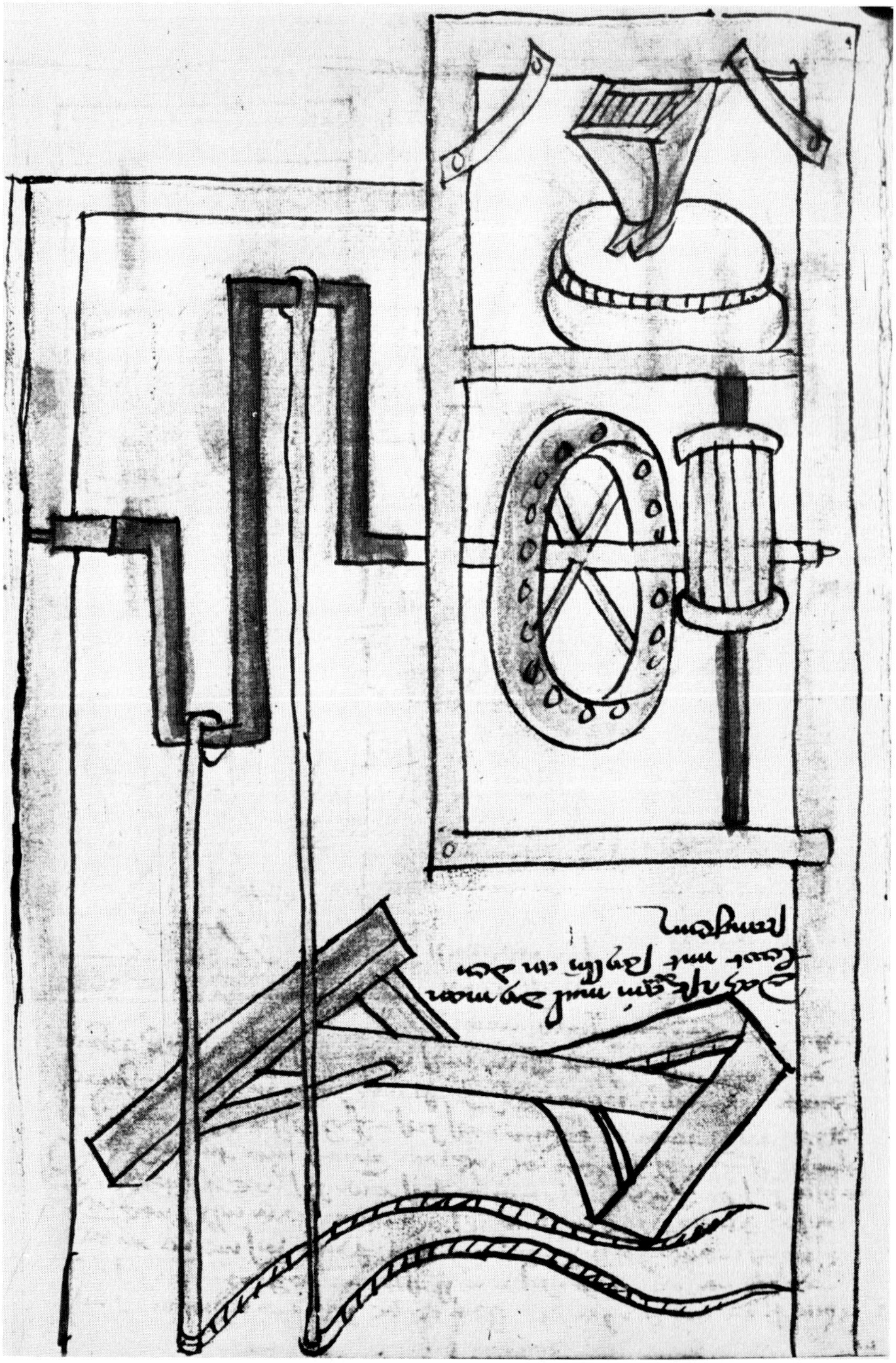

Ill. 2. A German technical drawing showing several problems in depicting complex devices before the advent of linear perspective. The connecting-rods attached to the crankshaft actually should project forward toward the viewer and rest atop the trestle shown below to form rocker-arms actuated by pulls on the rope. The text (which is inverted in the original) makes this somewhat clearer.

M:H. Hausbuch

Ill. 3. A slightly later German technical drawing showing pronounced Italian influence. Note how readily the relationship of all parts and actions can be perceived, including the complex bellows linkages in the upper right corner.

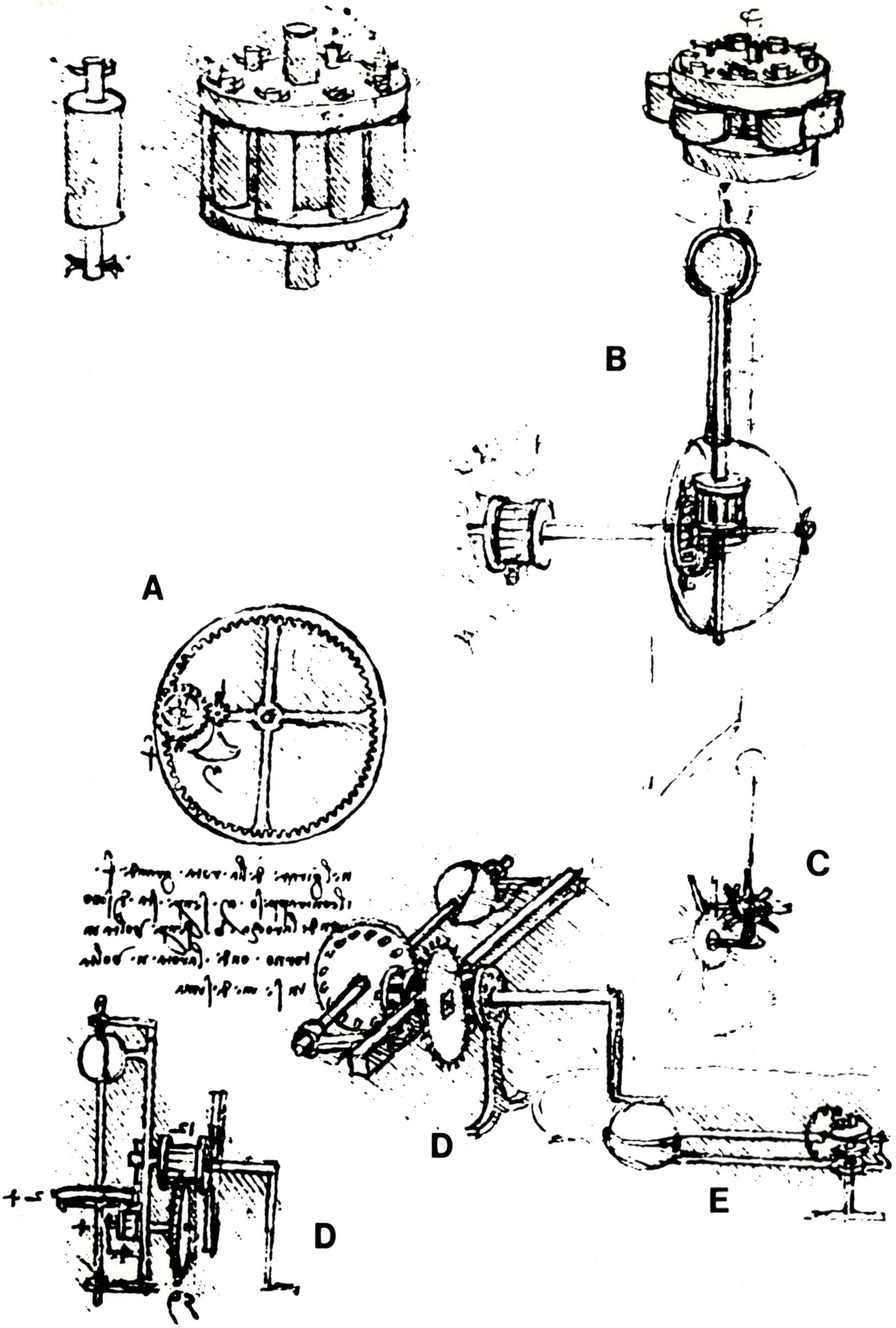

Ill. 4. Leonardo's sequence of sketches for a complex gear train in an astronomical clock. Sketches A-B show progressively more elaborate designs, each of which will work, but each of which suffers from some slight mechanical flaw.

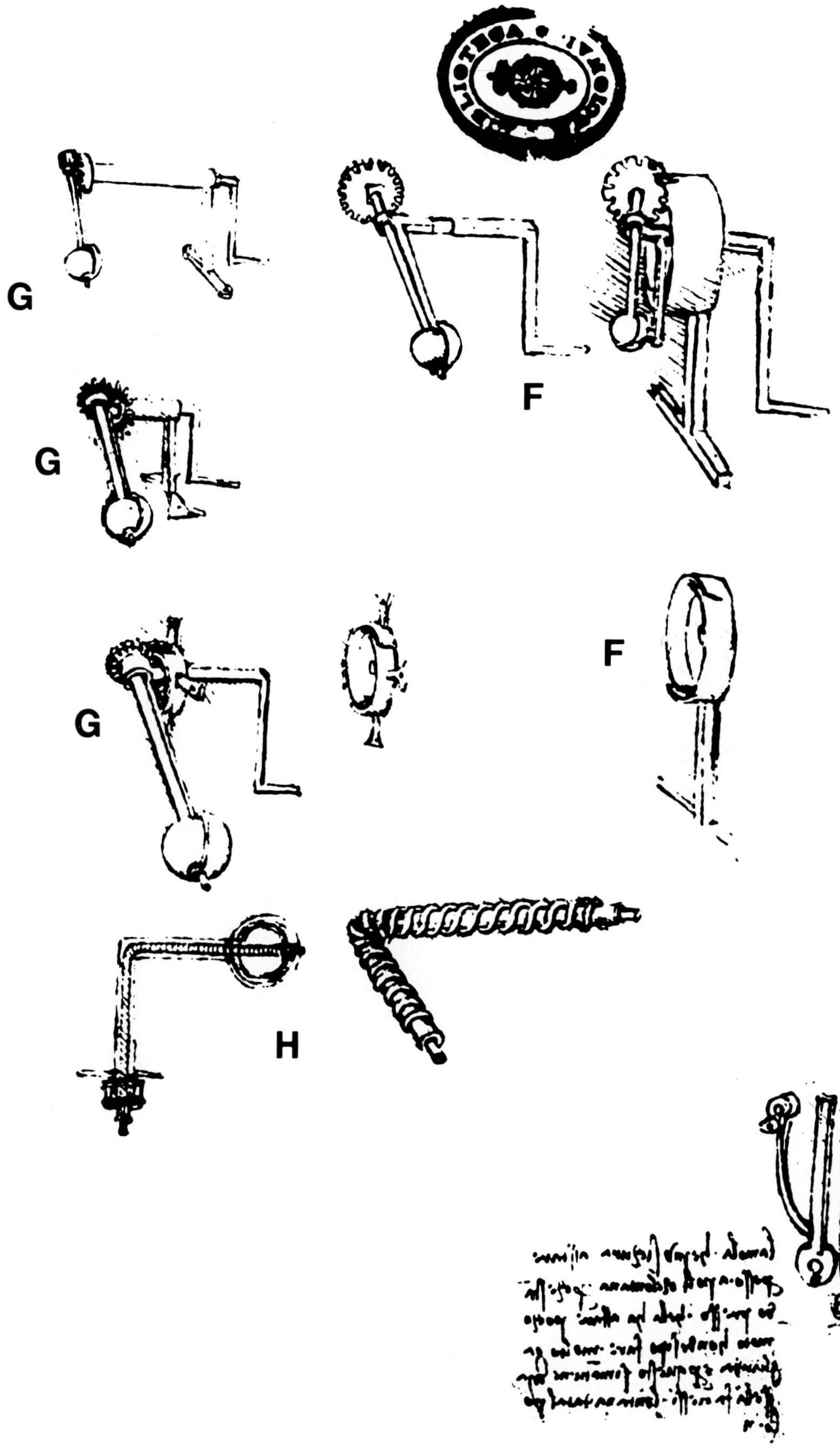

Ill. 5. The following page in the manuscript, which shows further development of Leonardo's designs toward version H, which was mechanically the simplest, but impossible in practice owing to limitations of materials in Leonardo's day.

Ill. 6. A representative drawing from Ramelli's work.
Note the precise rendering of detail possible with copperplate engraving
as well as the use of disassembled segments to show the relationship of various parts.

Ill. 1. Sloan Astrolabe.

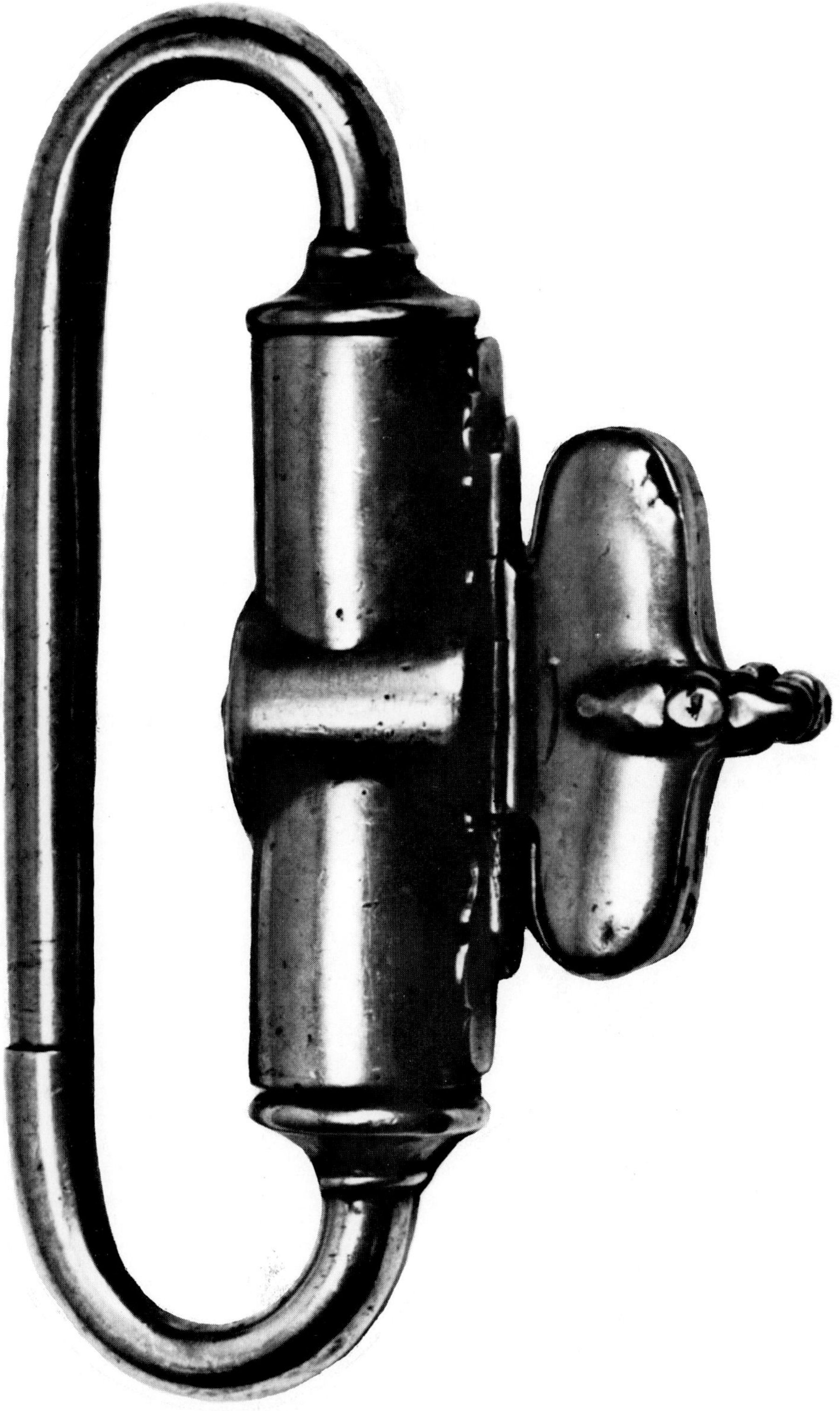

Ill. 1. Lock from a shrine in Isfahan, 15th-16th century.

Ill. 2. Brass lock from Isfahan, Shah Abbasi form, 17th century.

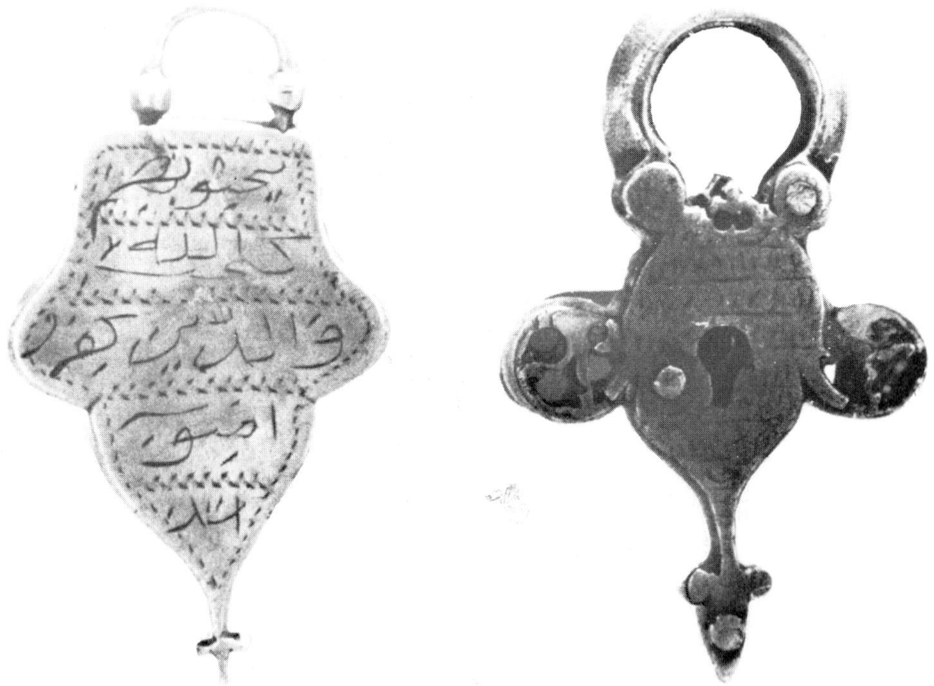

Ill. 3. Talisman locks, 18th and 19th centuries.

Ill. 4. Treasury lock, Safavi period, Iran, 16th century.

Ill. 5. Rabbit sculpture, Egypt, 11th century.

Ill. 6. Bronze lion sculpture with serpent tail, Egypt, 11th century.

Ill. 7. Antelope, Egypt, 11th century.

Ill. 8. Engraved silver lock,
18th or early 19th century.

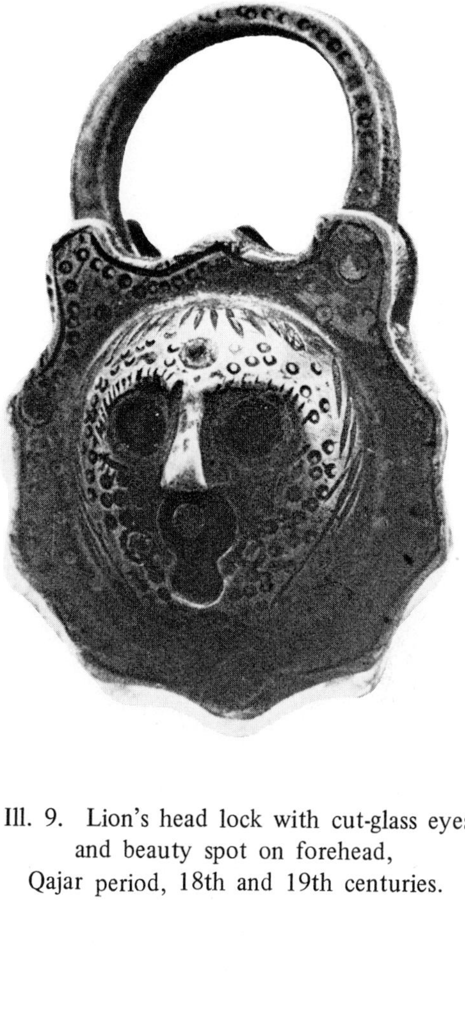

Ill. 9. Lion's head lock with cut-glass eyes
and beauty spot on forehead,
Qajar period, 18th and 19th centuries.

Ill. 10. Sir-Dar Madrasa of Samarqand,
17th century.

Ill. 11. Brass figural sand-cast lock, male figure in coat and breeches, probably 17th century.

Ill. 13. Shrine lock from Qajar period with false and genuine inscriptions.

Ill. 12. Tomb of Muslim princess from Nigde, Anatolia, 1321.